iRODS
User Group Meeting 2016
Proceedings

8TH ANNUAL CONFERENCE SUMMARY

The iRODS User Group Meeting of 2016 gathered together iRODS users, Consortium members, and staff to discuss iRODS-enabled applications and discoveries, technologies developed around iRODS, and future development and sustainability of iRODS and the iRODS Consortium.

The two-day event was held on June 8th and 9th in Chapel Hill, North Carolina, hosted by the iRODS Consortium, with over 90 people attending. Attendees and presenters represented over 30 academic, government, and commercial institutions.

Contents

Listing of Presentations.. 1

ARTICLES

iRODS Audit (C++) Rule Engine Plugin and AMQP 5
Terrell Russell, Jason Coposky, Justin James - RENCI at UNC Chapel Hill

Speed Research Discovery with Comprehensive Storage and Data Management 21
HGST | PANASAS | iRODS

Integrating HUBzero and iRODS 23
Rajesh Kalyanam, Robert A. Campbell, Samuel P. Wilson, Pascal Meunier, Lan Zhao, Elizabeth A. Hillery,
Carol Song - Purdue University

An R Package to Access iRODS Directly 31
Radovan Chytracek, Bernhard Sonderegger, Richard Coté - Nestlé Institute of Health Sciences

Davrods, an Apache WebDAV Interface to iRODS.. 41
Ton Smeele, Chris Smeele - Utrecht University

NFS-RODS: A Tool for Accessing iRODS Repositories via the NFS Protocol 49
D. Oliveira, A. Lobo Jr., F. Silva, G. Callou, I. Sousa, V. Alves, P. Maciel - UFPE
Stephen Worth - EMC Corporation
Jason Coposky - iRODS Consortium

Academic Workflow for Research Repositories Using iRODS and Object Storage 55
Randall Splinter - DDN

Application of iRODS Metadata Management for Cancer Genome Analysis Workflow 63
Lech Nieroda, Martin Peifer, Viktor Achter, Janna Velder, Ulrich Lang - University of Cologne

Status and Prospects of Kanki: An Open Source Cross-Platform Native iRODS Client Application 69
Ilari Korhonen, Miika Nurminen - University of Jyväskylä

Listing of Presentations

The following presentations were delivered at the meeting:

- **The iRODS Consortium in 2016**
  Jason Coposky, iRODS Consortium

- **iRODS 4.2 Overview**
  Terrell Russell, iRODS Consortium

- **Auditing with the Pluggable Rule Engine**
  Terrell Russell, iRODS Consortium

- **A Geo-Distributed Active Archive Tier**
  Earle Philhower, III, Western Digital

- **Testing Object Storage Systems with iRODS at Bayer**
  Othmar Weber, Bayer Business Systems

- **Advancing the Life Cycle of iRODS for Data**
  David Sallack, Panasas

- **Having it Both Ways: Bringing Data to Computation & Computation to Data with iRODS**
  Nirav Merchant, University of Arizona

- **Integrating HUBzero and iRODS**
  Rajesh Kalyanam, Purdue University

- **iRODS Data Integration with CloudyCluster Cloud-Based HPC**
  Boyd Wilson, Omnibond

- **Getting R to talk to iRODS**
  Bernhard Sonderegger, Nestlé Institute of Health Sciences

- **Davrods, an Apache WebDAV Interface to iRODS**
  Chris Smeele, Ton Smeele, Utrecht University

- **iRODS 4.3**
  Terrell Russell, iRODS Consortium

- **Bidirectional Integration of Multiple Metadata Sources**
  Hao Xu, DICE Group

- **DFC architecture & An iRODS Client for Mobile Devices**
  Jonathan Crabtree, Odum Institute
  Mike Conway, DICE Group
  Matthew Krause, DICE Group

- **NFS-RODS: A Tool for Accessing iRODS Repositories via the NFS Protocol**
  Danilo Oliveira, UFPE

- **MetaLnx: An Administrative and Metadata UI for iRODS**
  Stephen Worth, EMC

- **Academic Workflow for Research Repositories**
  Randy Splinter, DDN

- **Application of iRODS Metadata Management for Cancer Genome Analysis Workflow**
  Lech Nieroda, University of Cologne

- **Status and Prospects of Kanki: An Open Source Cross-Platform Native iRODS Client Application**
  Ilari Korhonen, University of Jyväskylä

- **DAM Secure File System**
  Paul Evans, Daystrom Technology Group

- **iRODS Feature Requests and Discussion**
  Reagan Moore, DICE Group

Articles

iRODS Audit (C++) Rule Engine Plugin and AMQP

Terrell Russell
Renaissance Computing
Institute (RENCI)
UNC Chapel Hill
unc@terrellrussell.com

Jason Coposky
Renaissance Computing
Institute (RENCI)
UNC Chapel Hill
jasonc@renci.org

Justin James
Renaissance Computing
Institute (RENCI)
UNC Chapel Hill
jjames@renci.org

ABSTRACT

iRODS 4.2 has introduced the new rule engine plugin interface. This interface offers the possibility of rule engines which support iRODS rules written in various languages. This paper introduces an audit plugin that emits a single AMQP message for every policy enforcement point within the iRODS server. We illustrate both the breadth and depth of these messages as well as some introductory analytics. This plugin may prove useful from instrumentation of a production iRODS installation to helping debug a confusing emergent distributed rule engine behavior.

Keywords

iRODS, audit, rule engine, automation, instrumentation, AMQP, dashboard

EXECUTION SEQUENCE OF POLICY ENFORCEMENT POINTS

By default, the audit plugin emits a single AMQP message per policy enforcement point hit within the iRODS server. This allows us to see the patterns inherent in the iRODS protocol as well as gain an understanding of the lifecycle of various client connections. The next few sections will contain complete policy enforcement point (PEP) execution sequences for `ils`, `imeta`, `iget`, `ireg`, `iput`, `iput` (1GiB large file), as well as an initial dashboard screenshot made possible by collecting the emitted AMQP messages.

Dynamic PEPs are listed in the following sections as `audit_pep_<plugin_operation>_<pre|post>`.

Legacy static PEPs are listed within parentheses and indented.

iRODS UGM 2016 June 8-9, 2016, Chapel Hill, NC

ils

When receiving an `ils` request, the server authenticates the incoming request, initiates a database connection, queries for the requested result set, and returns the results to the client.

Many dynamic policy enforcement points are hit multiple times as database and network traffic is sent and received.

174 dynamic PEPs + 4 static PEPs

```
audit_pep_network_read_header_pre
audit_pep_network_read_header_post
audit_pep_network_read_body_pre
audit_pep_network_read_body_post
audit_pep_database_open_pre
audit_pep_database_open_post
audit_pep_exec_rule_pre
    (acAclPolicy)
audit_pep_exec_microservice_pre
audit_pep_exec_microservice_post
audit_pep_exec_microservice_pre
audit_pep_database_gen_query_access_control_setup_pre
audit_pep_database_gen_query_access_control_setup_post
audit_pep_exec_microservice_post
audit_pep_exec_rule_post
audit_pep_database_gen_query_access_control_setup_pre
audit_pep_database_gen_query_access_control_setup_post
audit_pep_database_gen_query_pre
audit_pep_database_get_rcs_pre
audit_pep_database_get_rcs_post
audit_pep_database_gen_query_post
audit_pep_database_gen_query_access_control_setup_pre
audit_pep_database_gen_query_access_control_setup_post
audit_pep_database_gen_query_pre
audit_pep_database_get_rcs_pre
audit_pep_database_get_rcs_post
audit_pep_database_gen_query_post
audit_pep_exec_rule_pre
    (acChkHostAccessControl)
audit_pep_exec_microservice_pre
audit_pep_exec_microservice_post
audit_pep_exec_rule_post
audit_pep_exec_rule_pre
    (acSetPublicUserPolicy)
audit_pep_exec_microservice_pre
audit_pep_exec_microservice_post
audit_pep_exec_rule_post
audit_pep_exec_rule_pre
    (acPreConnect)
audit_pep_exec_microservice_pre
audit_pep_exec_microservice_post
audit_pep_exec_rule_post
audit_pep_network_write_body_pre
audit_pep_network_write_header_pre
audit_pep_network_write_header_post
audit_pep_network_write_body_post
audit_pep_network_agent_start_pre
audit_pep_network_agent_start_post
audit_pep_auth_agent_start_pre
audit_pep_auth_agent_start_post
audit_pep_network_read_header_pre
audit_pep_network_read_header_post
audit_pep_network_read_body_pre
audit_pep_network_read_body_post
audit_pep_auth_request_pre
audit_pep_auth_agent_auth_request_pre
audit_pep_auth_agent_auth_request_post
audit_pep_auth_request_post
audit_pep_network_write_body_pre
audit_pep_network_write_header_pre
audit_pep_network_write_header_post
audit_pep_network_write_body_post
audit_pep_auth_agent_start_pre
audit_pep_auth_agent_start_post
audit_pep_network_read_header_pre
audit_pep_network_read_header_post
audit_pep_network_read_body_pre
audit_pep_network_read_body_post
audit_pep_auth_response_pre
audit_pep_auth_agent_auth_response_pre

audit_pep_database_check_auth_pre
audit_pep_database_check_auth_post
audit_pep_auth_agent_auth_response_post
audit_pep_auth_response_post
audit_pep_network_write_body_pre
audit_pep_network_write_header_pre
audit_pep_network_write_header_post
audit_pep_network_write_body_post
audit_pep_auth_agent_start_pre
audit_pep_auth_agent_start_post
audit_pep_network_read_header_pre
audit_pep_network_read_header_post
audit_pep_network_read_body_pre
audit_pep_network_read_body_post
audit_pep_obj_stat_pre
audit_pep_database_gen_query_access_control_setup_pre
audit_pep_database_gen_query_access_control_setup_post
audit_pep_database_gen_query_pre
audit_pep_database_get_rcs_pre
audit_pep_database_get_rcs_post
audit_pep_database_gen_query_post
audit_pep_database_gen_query_access_control_setup_pre
audit_pep_database_gen_query_access_control_setup_post
audit_pep_database_gen_query_pre
audit_pep_database_get_rcs_pre
audit_pep_database_get_rcs_post
audit_pep_database_gen_query_post
audit_pep_database_gen_query_access_control_setup_pre
audit_pep_database_gen_query_access_control_setup_post
audit_pep_database_gen_query_pre
audit_pep_database_get_rcs_pre
audit_pep_database_get_rcs_post
audit_pep_database_gen_query_post
audit_pep_obj_stat_pre
audit_pep_network_write_body_pre
audit_pep_network_write_header_pre
audit_pep_network_write_header_post
audit_pep_network_write_body_post
audit_pep_auth_agent_start_pre
audit_pep_auth_agent_start_post
audit_pep_network_read_header_pre
audit_pep_network_read_header_post
audit_pep_network_read_body_pre
audit_pep_network_read_body_post
audit_pep_obj_stat_pre
audit_pep_database_gen_query_access_control_setup_pre
audit_pep_database_gen_query_access_control_setup_post
audit_pep_database_gen_query_pre
audit_pep_database_get_rcs_pre
audit_pep_database_get_rcs_post
audit_pep_database_gen_query_post
audit_pep_database_gen_query_access_control_setup_pre
audit_pep_database_gen_query_access_control_setup_post
audit_pep_database_gen_query_pre
audit_pep_database_get_rcs_pre
audit_pep_database_get_rcs_post
audit_pep_database_gen_query_post
audit_pep_obj_stat_post
audit_pep_network_write_body_pre
audit_pep_network_write_header_pre
audit_pep_network_write_header_post
audit_pep_network_write_body_post
audit_pep_auth_agent_start_pre
audit_pep_auth_agent_start_post
audit_pep_network_read_header_pre
audit_pep_network_read_header_post
audit_pep_network_read_body_pre
audit_pep_network_read_body_post
audit_pep_gen_query_pre
audit_pep_database_gen_query_access_control_setup_pre

audit_pep_database_gen_query_access_control_setup_post
audit_pep_database_gen_query_pre
audit_pep_database_get_rcs_pre
audit_pep_database_get_rcs_post
audit_pep_database_gen_query_post
audit_pep_gen_query_post
audit_pep_network_write_body_pre
audit_pep_network_write_header_pre
audit_pep_network_write_header_post
audit_pep_network_write_body_post
audit_pep_auth_agent_start_pre
audit_pep_auth_agent_start_post
audit_pep_network_read_header_pre
audit_pep_network_read_header_post
audit_pep_network_read_body_pre
audit_pep_network_read_body_post
audit_pep_gen_query_pre
audit_pep_database_gen_query_access_control_setup_pre
audit_pep_database_gen_query_access_control_setup_post
audit_pep_database_gen_query_pre
audit_pep_database_get_rcs_pre
audit_pep_database_get_rcs_post
audit_pep_database_gen_query_post
audit_pep_gen_query_post
audit_pep_network_write_body_pre
audit_pep_network_write_header_pre
audit_pep_network_write_header_post
audit_pep_network_write_body_post
audit_pep_auth_agent_start_pre
audit_pep_auth_agent_start_post
audit_pep_network_read_header_pre
audit_pep_network_read_header_post
audit_pep_network_read_body_pre
audit_pep_network_read_body_post
audit_pep_network_agent_stop_pre
audit_pep_network_agent_stop_post
audit_pep_database_close_pre
audit_pep_database_close_post
```

imeta

When receiving an `imeta` request, the server authenticates the incoming request, initiates a database connection, queries the metadata, and returns the results to the client.

106 dynamic PEPs + 6 static PEPs

```
audit_pep_network_read_header_pre
audit_pep_network_read_header_post
audit_pep_network_read_body_pre
audit_pep_network_read_body_post
audit_pep_database_open_pre
audit_pep_database_open_post
audit_pep_exec_rule_pre
    (acAclPolicy)
audit_pep_exec_microservice_pre
audit_pep_exec_microservice_post
audit_pep_exec_microservice_pre
audit_pep_database_gen_query_access_control_setup_pre
audit_pep_database_gen_query_access_control_setup_post
audit_pep_exec_microservice_post
audit_pep_exec_rule_post
audit_pep_database_gen_query_access_control_setup_pre
audit_pep_database_gen_query_access_control_setup_post
audit_pep_database_gen_query_pre
audit_pep_database_get_rcs_pre
audit_pep_database_get_rcs_post
audit_pep_database_gen_query_post
audit_pep_database_gen_query_access_control_setup_pre
audit_pep_database_gen_query_access_control_setup_post
audit_pep_database_gen_query_pre
audit_pep_database_get_rcs_pre
audit_pep_database_get_rcs_post
audit_pep_database_gen_query_post
audit_pep_exec_rule_pre
    (acChkHostAccessControl)
audit_pep_exec_microservice_pre
audit_pep_exec_microservice_post
audit_pep_exec_rule_post
audit_pep_exec_rule_pre
    (acSetPublicUserPolicy)
audit_pep_exec_microservice_pre
audit_pep_exec_microservice_post
audit_pep_exec_rule_post
audit_pep_exec_rule_pre
    (acPreConnect)
audit_pep_exec_microservice_pre
audit_pep_exec_microservice_post
audit_pep_exec_rule_post
audit_pep_network_write_body_pre
audit_pep_network_write_header_pre
audit_pep_network_write_header_post
audit_pep_network_write_body_post
audit_pep_network_agent_start_pre
audit_pep_network_agent_start_post
audit_pep_auth_agent_start_pre
audit_pep_auth_agent_start_post
audit_pep_network_read_header_pre
audit_pep_network_read_header_post
audit_pep_network_read_body_pre
audit_pep_network_read_body_post
audit_pep_auth_request_pre
audit_pep_auth_agent_auth_request_pre
audit_pep_auth_agent_auth_request_post
audit_pep_auth_request_post
audit_pep_network_write_body_pre
audit_pep_network_write_header_pre
audit_pep_network_write_header_post
audit_pep_network_write_body_post
audit_pep_auth_agent_start_pre
audit_pep_auth_agent_start_post
audit_pep_network_read_header_pre
audit_pep_network_read_header_post
audit_pep_network_read_body_pre
audit_pep_network_read_body_post
audit_pep_auth_response_pre
audit_pep_auth_agent_auth_response_pre
audit_pep_database_check_auth_pre
audit_pep_database_check_auth_post
audit_pep_auth_agent_auth_response_post
```

```
audit_pep_auth_response_post
audit_pep_network_write_body_pre
audit_pep_network_write_header_pre
audit_pep_network_write_header_post
audit_pep_network_write_body_post
audit_pep_auth_agent_start_pre
audit_pep_auth_agent_start_post
audit_pep_network_read_header_pre
audit_pep_network_read_header_post
audit_pep_network_read_body_pre
audit_pep_network_read_body_post
audit_pep_mod_avu_metadata_pre
audit_pep_exec_rule_pre
    (acPreProcForModifyAVUMetadata)
audit_pep_exec_microservice_pre
audit_pep_exec_microservice_post
audit_pep_exec_rule_post
audit_pep_database_add_avu_metadata_pre
audit_pep_database_add_avu_metadata_post
audit_pep_exec_rule_pre
    (acPostProcForModifyAVUMetadata)
audit_pep_exec_microservice_pre
audit_pep_exec_microservice_post
audit_pep_exec_rule_post
audit_pep_mod_avu_metadata_post
audit_pep_network_write_body_pre
audit_pep_network_write_header_pre
audit_pep_network_write_header_post
audit_pep_network_write_body_post
audit_pep_auth_agent_start_pre
audit_pep_auth_agent_start_post
audit_pep_network_read_header_pre
audit_pep_network_read_header_post
audit_pep_network_read_body_pre
audit_pep_network_read_body_post
audit_pep_network_agent_stop_pre
audit_pep_network_agent_stop_post
audit_pep_database_close_pre
audit_pep_database_close_post
```

iget

When receiving an `iget` request, the server authenticates the incoming request, initiates a database connection, checks permissions, opens, reads, and closes the file, and then sends the contents to the client.

148 dynamic PEPs + 6 static PEPs

```
audit_pep_network_read_header_pre          audit_pep_auth_response_post                         audit_pep_network_read_header_pre
audit_pep_network_read_header_post         audit_pep_network_write_body_pre                     audit_pep_network_read_header_post
audit_pep_network_read_body_pre            audit_pep_network_write_header_pre                   audit_pep_network_read_body_pre
audit_pep_network_read_body_post           audit_pep_network_write_header_post                  audit_pep_network_read_body_post
audit_pep_database_open_pre                audit_pep_network_write_body_post                    audit_pep_network_agent_stop_pre
audit_pep_database_open_post               audit_pep_auth_agent_start_pre                       audit_pep_network_agent_stop_post
audit_pep_exec_rule_pre                    audit_pep_auth_agent_start_post                      audit_pep_database_close_pre
    (acAclPolicy)                          audit_pep_network_read_header_pre                    audit_pep_database_close_post
audit_pep_exec_microservice_pre            audit_pep_network_read_header_post
audit_pep_exec_microservice_post           audit_pep_network_read_body_pre
audit_pep_exec_microservice_pre            audit_pep_network_read_body_post
audit_pep_database_gen_query_access_control_setup_pre     audit_pep_obj_stat_pre
audit_pep_database_gen_query_access_control_setup_post    audit_pep_database_gen_query_access_control_setup_pre
audit_pep_exec_microservice_post           audit_pep_database_gen_query_access_control_setup_post
audit_pep_exec_rule_post                   audit_pep_database_gen_query_pre
audit_pep_database_gen_query_access_control_setup_pre     audit_pep_database_get_rcs_pre
audit_pep_database_gen_query_access_control_setup_post    audit_pep_database_get_rcs_post
audit_pep_database_gen_query_pre           audit_pep_database_gen_query_post
audit_pep_database_get_rcs_pre             audit_pep_database_gen_query_access_control_setup_pre
audit_pep_database_get_rcs_post            audit_pep_database_gen_query_access_control_setup_post
audit_pep_database_gen_query_post          audit_pep_database_gen_query_pre
audit_pep_database_gen_query_access_control_setup_pre     audit_pep_database_get_rcs_pre
audit_pep_database_gen_query_access_control_setup_post    audit_pep_database_get_rcs_post
audit_pep_database_gen_query_pre           audit_pep_database_gen_query_post
audit_pep_database_get_rcs_pre             audit_pep_obj_stat_post
audit_pep_database_get_rcs_post            audit_pep_network_write_body_pre
audit_pep_database_gen_query_post          audit_pep_network_write_header_pre
audit_pep_exec_rule_pre                    audit_pep_network_write_header_post
    (acChkHostAccessControl)               audit_pep_network_write_body_post
audit_pep_exec_microservice_pre            audit_pep_auth_agent_start_pre
audit_pep_exec_microservice_post           audit_pep_auth_agent_start_post
audit_pep_exec_rule_post                   audit_pep_network_read_header_pre
audit_pep_exec_rule_pre                    audit_pep_network_read_header_post
    (acSetPublicUserPolicy)                audit_pep_network_read_body_pre
audit_pep_exec_microservice_pre            audit_pep_network_read_body_post
audit_pep_exec_microservice_post           audit_pep_data_obj_get_pre
audit_pep_exec_rule_post                   audit_pep_database_gen_query_access_control_setup_pre
audit_pep_exec_rule_pre                    audit_pep_database_gen_query_access_control_setup_post
    (acPreConnect)                         audit_pep_database_gen_query_pre
audit_pep_exec_microservice_pre            audit_pep_database_get_rcs_pre
audit_pep_exec_microservice_post           audit_pep_database_get_rcs_post
audit_pep_exec_rule_post                   audit_pep_database_gen_query_post
audit_pep_network_write_body_pre           audit_pep_database_gen_query_access_control_setup_pre
audit_pep_network_write_header_pre         audit_pep_database_gen_query_access_control_setup_post
audit_pep_network_write_header_post        audit_pep_database_gen_query_pre
audit_pep_network_write_body_post          audit_pep_database_get_rcs_pre
audit_pep_network_agent_start_pre          audit_pep_database_get_rcs_post
audit_pep_network_agent_start_post         audit_pep_database_gen_query_post
audit_pep_auth_agent_start_pre             audit_pep_resource_resolve_hierarchy_pre
audit_pep_auth_agent_start_post            audit_pep_resource_resolve_hierarchy_post
audit_pep_network_read_header_pre          audit_pep_exec_rule_pre
audit_pep_network_read_header_post             (acPreprocForDataObjOpen)
audit_pep_network_read_body_pre            audit_pep_exec_microservice_pre
audit_pep_network_read_body_post           audit_pep_exec_microservice_post
audit_pep_auth_request_pre                 audit_pep_exec_rule_post
audit_pep_auth_agent_auth_request_pre      audit_pep_resource_open_pre
audit_pep_auth_agent_auth_request_post     audit_pep_resource_open_post
audit_pep_auth_request_post                audit_pep_resource_read_pre
audit_pep_network_write_body_pre           audit_pep_resource_read_post
audit_pep_network_write_header_pre         audit_pep_resource_close_pre
audit_pep_network_write_header_post        audit_pep_resource_close_post
audit_pep_network_write_body_post          audit_pep_exec_rule_pre
audit_pep_auth_agent_start_pre                 (acPostProcForOpen)
audit_pep_auth_agent_start_post            audit_pep_exec_microservice_pre
audit_pep_network_read_header_pre          audit_pep_exec_microservice_post
audit_pep_network_read_header_post         audit_pep_exec_rule_post
audit_pep_network_read_body_pre            audit_pep_data_obj_get_post
audit_pep_network_read_body_post           audit_pep_network_write_body_pre
audit_pep_auth_response_pre                audit_pep_network_write_header_pre
audit_pep_auth_agent_auth_response_pre     audit_pep_network_write_header_post
audit_pep_database_check_auth_pre          audit_pep_network_write_body_post
audit_pep_database_check_auth_post         audit_pep_auth_agent_start_pre
audit_pep_auth_agent_auth_response_post    audit_pep_auth_agent_start_post
```

ireg

When receiving an **ireg** request, the server authenticates the incoming request, initiates a database connection, checks permissions, registers the new file, and checks quota usage.

168 dynamic PEPs + 7 static PEPs

```
audit_pep_network_read_header_pre
audit_pep_network_read_header_post
audit_pep_network_read_body_pre
audit_pep_network_read_body_post
audit_pep_database_open_pre
audit_pep_database_open_post
audit_pep_exec_rule_pre
     (acAclPolicy)
audit_pep_exec_microservice_pre
audit_pep_exec_microservice_post
audit_pep_exec_microservice_pre
audit_pep_database_gen_query_access_control_setup_pre
audit_pep_database_gen_query_access_control_setup_post
audit_pep_exec_microservice_post
audit_pep_exec_rule_post
audit_pep_database_gen_query_access_control_setup_pre
audit_pep_database_gen_query_access_control_setup_post
audit_pep_database_gen_query_pre
audit_pep_database_get_rcs_pre
audit_pep_database_get_rcs_post
audit_pep_database_gen_query_post
audit_pep_database_gen_query_access_control_setup_pre
audit_pep_database_gen_query_access_control_setup_post
audit_pep_database_gen_query_pre
audit_pep_database_get_rcs_pre
audit_pep_database_get_rcs_post
audit_pep_database_gen_query_post
audit_pep_exec_rule_pre
     (acChkHostAccessControl)
audit_pep_exec_microservice_pre
audit_pep_exec_microservice_post
audit_pep_exec_rule_post
audit_pep_exec_rule_pre
     (acSetPublicUserPolicy)
audit_pep_exec_microservice_pre
audit_pep_exec_microservice_post
audit_pep_exec_rule_post
audit_pep_exec_rule_pre
     (acPreConnect)
audit_pep_exec_microservice_pre
audit_pep_exec_microservice_post
audit_pep_exec_rule_post
audit_pep_network_write_body_pre
audit_pep_network_write_header_pre
audit_pep_network_write_header_post
audit_pep_network_write_body_post
audit_pep_network_agent_start_pre
audit_pep_network_agent_start_post
audit_pep_auth_agent_start_pre
audit_pep_auth_agent_start_post
audit_pep_network_read_header_pre
audit_pep_network_read_header_post
audit_pep_network_read_body_pre
audit_pep_network_read_body_post
audit_pep_auth_request_pre
audit_pep_auth_agent_auth_request_pre
audit_pep_auth_agent_auth_request_post
audit_pep_auth_request_post
audit_pep_network_write_body_pre
audit_pep_network_write_header_pre
audit_pep_network_write_header_post
audit_pep_network_write_body_post
audit_pep_auth_agent_start_pre
audit_pep_auth_agent_start_post
audit_pep_network_read_header_pre
audit_pep_network_read_header_post
audit_pep_network_read_body_pre
audit_pep_network_read_body_post
audit_pep_auth_response_pre
audit_pep_auth_agent_auth_response_pre
audit_pep_database_check_auth_pre
audit_pep_database_check_auth_post
audit_pep_auth_agent_auth_response_post

audit_pep_auth_response_post
audit_pep_network_write_body_pre
audit_pep_network_write_header_pre
audit_pep_network_write_header_post
audit_pep_network_write_body_post
audit_pep_auth_agent_start_pre
audit_pep_auth_agent_start_post
audit_pep_network_read_header_pre
audit_pep_network_read_header_post
audit_pep_network_read_body_pre
audit_pep_network_read_body_post
audit_pep_obj_stat_pre
audit_pep_database_gen_query_access_control_setup_pre
audit_pep_database_gen_query_access_control_setup_post
audit_pep_database_gen_query_pre
audit_pep_database_get_rcs_pre
audit_pep_database_get_rcs_post
audit_pep_database_gen_query_post
audit_pep_database_gen_query_access_control_setup_pre
audit_pep_database_gen_query_access_control_setup_post
audit_pep_database_gen_query_pre
audit_pep_database_get_rcs_pre
audit_pep_database_get_rcs_post
audit_pep_database_gen_query_post
audit_pep_database_gen_query_access_control_setup_pre
audit_pep_database_gen_query_access_control_setup_post
audit_pep_database_gen_query_pre
audit_pep_database_get_rcs_pre
audit_pep_database_get_rcs_post
audit_pep_database_gen_query_post
audit_pep_obj_stat_post
audit_pep_network_write_body_pre
audit_pep_network_write_header_pre
audit_pep_network_write_header_post
audit_pep_network_write_body_post
audit_pep_auth_agent_start_pre
audit_pep_auth_agent_start_post
audit_pep_network_read_header_pre
audit_pep_network_read_header_post
audit_pep_network_read_body_pre
audit_pep_network_read_body_post
audit_pep_phy_path_reg_pre
audit_pep_database_gen_query_access_control_setup_pre
audit_pep_database_gen_query_access_control_setup_post
audit_pep_database_gen_query_pre
audit_pep_database_get_rcs_pre
audit_pep_database_get_rcs_post
audit_pep_database_gen_query_post
audit_pep_database_gen_query_access_control_setup_pre
audit_pep_database_gen_query_access_control_setup_post
audit_pep_database_gen_query_pre
audit_pep_database_get_rcs_pre
audit_pep_database_get_rcs_post
audit_pep_database_gen_query_post
audit_pep_exec_rule_pre
     (acSetRescSchemeForCreate)
audit_pep_exec_microservice_pre
audit_pep_exec_microservice_post
audit_pep_exec_microservice_pre
audit_pep_exec_microservice_post
audit_pep_exec_rule_post
audit_pep_exec_rule_pre
     (acRescQuotaPolicy)
audit_pep_exec_microservice_pre
audit_pep_exec_microservice_post
audit_pep_exec_microservice_pre
audit_pep_exec_microservice_post

audit_pep_exec_rule_post
audit_pep_resource_resolve_hierarchy_pre
audit_pep_resource_resolve_hierarchy_post
audit_pep_resource_stat_pre
audit_pep_resource_stat_post
audit_pep_database_reg_data_obj_pre
audit_pep_database_reg_data_obj_post
audit_pep_resource_registered_pre
audit_pep_resource_registered_post
audit_pep_exec_rule_pre
     (acPostProcForFilePathReg)
audit_pep_exec_microservice_pre
audit_pep_exec_microservice_post
audit_pep_exec_rule_post
audit_pep_phy_path_reg_post
audit_pep_network_write_body_pre
audit_pep_network_write_header_pre
audit_pep_network_write_header_post
audit_pep_network_write_body_post
audit_pep_auth_agent_start_pre
audit_pep_auth_agent_start_post
audit_pep_network_read_header_pre
audit_pep_network_read_header_post
audit_pep_network_read_body_pre
audit_pep_network_read_body_post
audit_pep_network_agent_stop_pre
audit_pep_network_agent_stop_post
audit_pep_database_close_pre
audit_pep_database_close_post
```

iput

When receiving an **iput** request of a small file (by default, smaller than 32MiB), the server authenticates the incoming request, initiates a database connection, checks permissions, receives and writes the incoming file to disk, registers the new file, and checks quota usage.

234 dynamic PEPs + 11 static PEPs

```
audit_pep_network_read_header_pre
audit_pep_network_read_header_post
audit_pep_network_read_body_pre
audit_pep_network_read_body_post
audit_pep_database_open_pre
audit_pep_database_open_post
audit_pep_exec_rule_pre
      (acAclPolicy)
audit_pep_exec_microservice_pre
audit_pep_exec_microservice_post
audit_pep_exec_microservice_pre
audit_pep_database_gen_query_access_control_setup_pre
audit_pep_database_gen_query_access_control_setup_post
audit_pep_exec_microservice_post
audit_pep_exec_rule_post
audit_pep_database_gen_query_access_control_setup_pre
audit_pep_database_gen_query_access_control_setup_post
audit_pep_database_gen_query_pre
audit_pep_database_get_rcs_pre
audit_pep_database_get_rcs_post
audit_pep_database_gen_query_post
audit_pep_database_gen_query_access_control_setup_pre
audit_pep_database_gen_query_access_control_setup_post
audit_pep_database_gen_query_pre
audit_pep_database_get_rcs_pre
audit_pep_database_get_rcs_post
audit_pep_database_gen_query_post
audit_pep_exec_rule_pre
      (acChkHostAccessControl)
audit_pep_exec_microservice_pre
audit_pep_exec_microservice_post
audit_pep_exec_rule_post
audit_pep_exec_rule_pre
      (acSetPublicUserPolicy)
audit_pep_exec_microservice_pre
audit_pep_exec_microservice_post
audit_pep_exec_rule_post
audit_pep_exec_rule_pre
      (acPreConnect)
audit_pep_exec_microservice_pre
audit_pep_exec_microservice_post
audit_pep_exec_rule_post
audit_pep_network_write_body_pre
audit_pep_network_write_header_pre
audit_pep_network_write_header_post
audit_pep_network_write_body_post
audit_pep_network_agent_start_pre
audit_pep_network_agent_start_post
audit_pep_auth_agent_start_pre
audit_pep_auth_agent_start_post
audit_pep_network_read_header_pre
audit_pep_network_read_header_post
audit_pep_network_read_body_pre
audit_pep_network_read_body_post
audit_pep_auth_request_pre
audit_pep_auth_agent_auth_request_pre
audit_pep_auth_agent_auth_request_post
audit_pep_auth_request_post
audit_pep_network_write_body_pre
audit_pep_network_write_header_pre
audit_pep_network_write_header_post
audit_pep_network_write_body_post
audit_pep_auth_agent_start_pre
audit_pep_auth_agent_start_post
audit_pep_network_read_header_pre
audit_pep_network_read_header_post
audit_pep_network_read_body_pre
audit_pep_network_read_body_post
audit_pep_auth_response_pre
audit_pep_auth_agent_auth_response_pre
audit_pep_database_check_auth_pre
audit_pep_database_check_auth_post
```

```
audit_pep_auth_agent_auth_response_post
audit_pep_auth_response_post
audit_pep_network_write_body_pre
audit_pep_network_write_header_pre
audit_pep_network_write_header_post
audit_pep_network_write_body_post
audit_pep_auth_agent_start_pre
audit_pep_auth_agent_start_post
audit_pep_network_read_header_pre
audit_pep_network_read_header_post
audit_pep_network_read_body_pre
audit_pep_network_read_body_post
audit_pep_obj_stat_pre
audit_pep_database_gen_query_access_control_setup_pre
audit_pep_database_gen_query_access_control_setup_post
audit_pep_database_gen_query_pre
audit_pep_database_get_rcs_pre
audit_pep_database_get_rcs_post
audit_pep_database_gen_query_post
audit_pep_database_gen_query_access_control_setup_pre
audit_pep_database_gen_query_access_control_setup_post
audit_pep_database_gen_query_pre
audit_pep_database_get_rcs_pre
audit_pep_database_get_rcs_post
audit_pep_database_gen_query_post
audit_pep_database_gen_query_access_control_setup_pre
audit_pep_database_gen_query_access_control_setup_post
audit_pep_database_gen_query_pre
audit_pep_database_get_rcs_pre
audit_pep_database_get_rcs_post
audit_pep_database_gen_query_post
audit_pep_obj_stat_post
audit_pep_network_write_body_pre
audit_pep_network_write_header_pre
audit_pep_network_write_header_post
audit_pep_network_write_body_post
audit_pep_auth_agent_start_pre
audit_pep_auth_agent_start_post
audit_pep_network_read_header_pre
audit_pep_network_read_header_post
audit_pep_network_read_body_pre
audit_pep_network_read_body_post
audit_pep_data_obj_put_pre
audit_pep_database_gen_query_access_control_setup_pre
audit_pep_database_gen_query_access_control_setup_post
audit_pep_database_gen_query_pre
audit_pep_database_get_rcs_pre
audit_pep_database_get_rcs_post
```

```
audit_pep_database_gen_query_post
audit_pep_database_gen_query_access_control_setup_pre
audit_pep_database_gen_query_access_control_setup_post
audit_pep_database_gen_query_pre
audit_pep_database_get_rcs_pre
audit_pep_database_get_rcs_post
audit_pep_database_gen_query_post
audit_pep_database_gen_query_access_control_setup_pre
audit_pep_database_gen_query_access_control_setup_post
audit_pep_database_gen_query_pre
audit_pep_database_get_rcs_pre
audit_pep_database_get_rcs_post
audit_pep_database_gen_query_post
audit_pep_database_gen_query_access_control_setup_pre
audit_pep_database_gen_query_access_control_setup_post
audit_pep_database_gen_query_pre
audit_pep_database_get_rcs_pre
audit_pep_database_get_rcs_post
audit_pep_database_gen_query_post
audit_pep_exec_rule_pre
      (acSetRescSchemeForCreate)
audit_pep_exec_microservice_pre
audit_pep_exec_microservice_post
audit_pep_exec_microservice_pre
audit_pep_exec_microservice_post
audit_pep_exec_rule_post
audit_pep_exec_rule_pre
      (acRescQuotaPolicy)
audit_pep_exec_microservice_pre
audit_pep_exec_microservice_post
audit_pep_exec_microservice_pre
audit_pep_exec_microservice_post
audit_pep_exec_rule_post
audit_pep_resource_resolve_hierarchy_pre
audit_pep_resource_resolve_hierarchy_post
audit_pep_database_gen_query_access_control_setup_pre
audit_pep_database_gen_query_access_control_setup_post
audit_pep_database_gen_query_pre
audit_pep_database_get_rcs_pre
audit_pep_database_get_rcs_post
audit_pep_database_gen_query_post
audit_pep_exec_rule_pre
      (acSetRescSchemeForCreate)
audit_pep_exec_microservice_pre
audit_pep_exec_microservice_post
audit_pep_exec_microservice_pre
audit_pep_exec_microservice_post
audit_pep_exec_rule_post
audit_pep_exec_rule_pre
      (acSetVaultPathPolicy)
audit_pep_exec_microservice_pre
audit_pep_exec_microservice_post
audit_pep_exec_microservice_pre
audit_pep_exec_microservice_post
audit_pep_exec_rule_post
audit_pep_resource_create_pre
audit_pep_resource_create_post
audit_pep_resource_write_pre
audit_pep_resource_write_post
audit_pep_resource_close_pre
audit_pep_resource_close_post
audit_pep_database_reg_data_obj_pre
audit_pep_database_reg_data_obj_post
audit_pep_resource_registered_pre
audit_pep_resource_registered_post
audit_pep_resource_stat_pre
audit_pep_resource_stat_post
audit_pep_exec_rule_pre
      (acPreProcForModifyDataObjMeta)
audit_pep_exec_microservice_pre
audit_pep_exec_microservice_post
audit_pep_exec_rule_post
```

```
audit_pep_database_mod_data_obj_meta_pre
audit_pep_database_mod_data_obj_meta_post
audit_pep_exec_rule_pre
      (acPostProcForModifyDataObjMeta)
audit_pep_exec_microservice_pre
audit_pep_exec_microservice_post
audit_pep_exec_rule_post
audit_pep_resource_modified_pre
audit_pep_resource_modified_post
audit_pep_exec_rule_pre
      (acPostProcForPut)
audit_pep_exec_microservice_pre
audit_pep_exec_microservice_post
audit_pep_exec_rule_post
audit_pep_data_obj_put_post
audit_pep_network_write_body_pre
audit_pep_network_write_header_pre
audit_pep_network_write_header_post
audit_pep_network_write_body_post
audit_pep_auth_agent_start_pre
audit_pep_auth_agent_start_post
audit_pep_network_read_header_pre
audit_pep_network_read_header_post
audit_pep_network_read_body_pre
audit_pep_network_read_body_post
audit_pep_network_agent_stop_pre
audit_pep_network_agent_stop_post
audit_pep_database_close_pre
audit_pep_database_close_post
```

iput (1GiB large file)

When receiving an `iput` request of a large file (by default, 32MiB or larger), the server authenticates the incoming request, initiates a database connection, checks permissions, registers a zero length file as placeholder, receives and writes the incoming parallel file transfer to disk piece by piece, updates the registered file in the catalog, and checks quota usage.

978 dynamic PEPs + 44 static PEPs

```
audit_pep_network_read_header_pre
audit_pep_network_read_header_post
audit_pep_network_read_body_pre
audit_pep_network_read_body_post
audit_pep_database_open_pre
audit_pep_database_open_post
audit_pep_exec_rule_pre
    (acAclPolicy)
audit_pep_exec_microservice_pre
audit_pep_exec_microservice_post
audit_pep_exec_microservice_pre
audit_pep_database_gen_query_access_control_setup_pre
audit_pep_database_gen_query_access_control_setup_post
audit_pep_exec_microservice_post
audit_pep_exec_rule_post
audit_pep_database_gen_query_access_control_setup_pre
audit_pep_database_gen_query_access_control_setup_post
audit_pep_database_gen_query_pre
audit_pep_database_get_rcs_pre
audit_pep_database_get_rcs_post
audit_pep_database_gen_query_post
audit_pep_database_gen_query_access_control_setup_pre
audit_pep_database_gen_query_access_control_setup_post
audit_pep_database_gen_query_pre
audit_pep_database_get_rcs_pre
audit_pep_database_get_rcs_post
audit_pep_database_gen_query_post
audit_pep_exec_rule_pre
    (acChkHostAccessControl)
audit_pep_exec_microservice_pre
audit_pep_exec_microservice_post
audit_pep_exec_rule_post
audit_pep_exec_rule_pre
    (acSetPublicUserPolicy)
audit_pep_exec_microservice_pre
audit_pep_exec_microservice_post
audit_pep_exec_rule_post
audit_pep_exec_rule_pre
    (acPreConnect)
audit_pep_exec_microservice_pre
audit_pep_exec_microservice_post
audit_pep_exec_rule_post
audit_pep_network_write_body_pre
audit_pep_network_write_header_pre
audit_pep_network_write_header_post
audit_pep_network_write_body_post
audit_pep_network_agent_start_pre
audit_pep_network_agent_start_post
audit_pep_auth_agent_start_pre
audit_pep_auth_agent_start_post
audit_pep_network_read_header_pre
audit_pep_network_read_header_post
audit_pep_network_read_body_pre
audit_pep_network_read_body_post
audit_pep_auth_request_pre
audit_pep_auth_agent_auth_request_pre
audit_pep_auth_agent_auth_request_post
audit_pep_auth_request_post
audit_pep_network_write_body_pre
audit_pep_network_write_header_pre
audit_pep_network_write_header_post
audit_pep_network_write_body_post
audit_pep_auth_agent_start_pre
audit_pep_auth_agent_start_post
audit_pep_network_read_header_pre
audit_pep_network_read_header_post
audit_pep_network_read_body_pre
audit_pep_network_read_body_post
audit_pep_auth_response_pre
audit_pep_auth_agent_auth_response_pre

audit_pep_database_check_auth_pre
audit_pep_database_check_auth_post
audit_pep_auth_agent_auth_response_post
audit_pep_auth_response_post
audit_pep_network_write_body_pre
audit_pep_network_write_header_pre
audit_pep_network_write_body_post
audit_pep_auth_agent_start_pre
audit_pep_auth_agent_start_post
audit_pep_network_read_header_pre
audit_pep_network_read_header_post
audit_pep_network_read_body_pre
audit_pep_network_read_body_post
audit_pep_obj_stat_pre
audit_pep_database_gen_query_access_control_setup_pre
audit_pep_database_gen_query_access_control_setup_post
audit_pep_database_gen_query_pre
audit_pep_database_get_rcs_pre
audit_pep_database_get_rcs_post
audit_pep_database_gen_query_post
audit_pep_database_gen_query_access_control_setup_pre
audit_pep_database_gen_query_access_control_setup_post
audit_pep_database_gen_query_pre
audit_pep_database_get_rcs_pre
audit_pep_database_get_rcs_post
audit_pep_database_gen_query_post
audit_pep_obj_stat_post
audit_pep_network_write_body_pre
audit_pep_network_write_header_pre
audit_pep_network_write_header_post
audit_pep_network_write_body_post
audit_pep_auth_agent_start_pre
audit_pep_auth_agent_start_post
audit_pep_network_read_header_pre
audit_pep_network_read_header_post
audit_pep_network_read_body_pre
audit_pep_network_read_body_post
audit_pep_obj_stat_pre
audit_pep_database_gen_query_access_control_setup_pre
audit_pep_database_gen_query_access_control_setup_post
audit_pep_database_gen_query_pre
audit_pep_database_get_rcs_pre
audit_pep_database_get_rcs_post
audit_pep_database_gen_query_post
audit_pep_database_gen_query_access_control_setup_pre
audit_pep_database_gen_query_access_control_setup_post
audit_pep_database_gen_query_pre
audit_pep_database_get_rcs_pre
audit_pep_database_get_rcs_post
audit_pep_database_gen_query_post
audit_pep_obj_stat_post
audit_pep_network_write_body_pre
audit_pep_network_write_header_pre
audit_pep_network_write_header_post
audit_pep_network_write_body_post
audit_pep_auth_agent_start_pre
audit_pep_auth_agent_start_post
audit_pep_network_read_header_pre
audit_pep_network_read_header_post
audit_pep_network_read_body_pre
audit_pep_network_read_body_post
audit_pep_data_obj_put_pre
audit_pep_database_gen_query_access_control_setup_pre

audit_pep_database_gen_query_access_control_setup_post
audit_pep_database_gen_query_pre
audit_pep_database_get_rcs_pre
audit_pep_database_get_rcs_post
audit_pep_database_gen_query_post
audit_pep_database_gen_query_access_control_setup_pre
audit_pep_database_gen_query_access_control_setup_post
audit_pep_database_gen_query_pre
audit_pep_database_get_rcs_pre
audit_pep_database_get_rcs_post
audit_pep_database_gen_query_post
audit_pep_database_gen_query_access_control_setup_pre
audit_pep_database_gen_query_access_control_setup_post
audit_pep_database_gen_query_pre
audit_pep_database_get_rcs_pre
audit_pep_database_get_rcs_post
audit_pep_database_gen_query_post
audit_pep_database_gen_query_access_control_setup_pre
audit_pep_database_gen_query_access_control_setup_post
audit_pep_database_gen_query_pre
audit_pep_database_get_rcs_pre
audit_pep_database_get_rcs_post
audit_pep_database_gen_query_post
audit_pep_exec_rule_pre
    (acSetRescSchemeForCreate)
audit_pep_exec_microservice_pre
audit_pep_exec_microservice_post
audit_pep_exec_microservice_pre
audit_pep_exec_microservice_post
audit_pep_exec_rule_post
audit_pep_exec_rule_pre
    (acRescQuotaPolicy)
audit_pep_exec_microservice_pre
audit_pep_exec_microservice_post
audit_pep_exec_microservice_pre
audit_pep_exec_microservice_post
audit_pep_exec_rule_post
audit_pep_resource_resolve_hierarchy_pre
audit_pep_resource_resolve_hierarchy_post
audit_pep_database_gen_query_access_control_setup_pre
audit_pep_database_gen_query_access_control_setup_post
audit_pep_database_gen_query_pre
audit_pep_database_get_rcs_pre
audit_pep_database_get_rcs_post
audit_pep_database_gen_query_post
audit_pep_exec_rule_pre
    (acSetRescSchemeForCreate)
audit_pep_exec_microservice_pre
audit_pep_exec_microservice_post
audit_pep_exec_microservice_pre
audit_pep_exec_microservice_post
audit_pep_exec_rule_post
audit_pep_exec_rule_pre
    (acSetVaultPathPolicy)
audit_pep_exec_microservice_pre
audit_pep_exec_microservice_post
audit_pep_exec_microservice_pre
audit_pep_exec_microservice_post
audit_pep_exec_rule_post
audit_pep_resource_create_pre
audit_pep_resource_create_post
audit_pep_database_reg_data_obj_pre
audit_pep_database_reg_data_obj_post
audit_pep_resource_registered_pre
audit_pep_resource_registered_post
audit_pep_exec_rule_pre
    (acSetNumThreads)
audit_pep_exec_microservice_pre
audit_pep_exec_microservice_post
audit_pep_exec_microservice_pre
```

```
audit_pep_exec_microservice_post          audit_pep_resource_open_pre               audit_pep_resource_write_pre
audit_pep_exec_rule_post                  audit_pep_resource_write_pre              audit_pep_resource_open_pre
audit_pep_network_write_body_pre          audit_pep_resource_write_post             audit_pep_resource_write_post
audit_pep_network_write_header_pre        audit_pep_resource_write_pre              audit_pep_resource_write_pre
audit_pep_network_write_header_post       audit_pep_resource_write_post             audit_pep_resource_write_post
audit_pep_network_write_body_post         audit_pep_resource_write_pre              audit_pep_resource_write_pre
audit_pep_exec_rule_pre                   audit_pep_resource_write_pre              audit_pep_resource_write_post
    (acPreProcForServerPortal)            audit_pep_resource_write_post             audit_pep_resource_write_pre
audit_pep_exec_microservice_pre           audit_pep_resource_write_pre              audit_pep_resource_write_post
audit_pep_exec_microservice_post          audit_pep_resource_open_post              audit_pep_resource_write_pre
audit_pep_exec_rule_post                  audit_pep_resource_write_post             audit_pep_resource_write_post
audit_pep_exec_rule_pre                   audit_pep_exec_rule_pre                   audit_pep_resource_write_pre
    (acPreProcForServerPortal)                (acPreProcForServerPortal)            audit_pep_resource_write_post
audit_pep_exec_microservice_pre           audit_pep_exec_microservice_pre           audit_pep_resource_write_pre
audit_pep_exec_microservice_post          audit_pep_exec_microservice_post          audit_pep_resource_write_post
audit_pep_exec_rule_post                  audit_pep_resource_lseek_pre              audit_pep_resource_write_pre
audit_pep_resource_open_pre               audit_pep_exec_rule_post                  audit_pep_resource_write_post
audit_pep_resource_open_post              audit_pep_resource_lseek_post             audit_pep_resource_write_pre
audit_pep_exec_rule_pre                   audit_pep_resource_open_pre               audit_pep_resource_write_post
    (acPreProcForServerPortal)            audit_pep_resource_write_pre              audit_pep_resource_open_post
audit_pep_exec_microservice_pre           audit_pep_resource_write_post             audit_pep_exec_rule_pre
audit_pep_exec_microservice_post          audit_pep_resource_write_pre                  (acPreProcForServerPortal)
audit_pep_resource_lseek_pre              audit_pep_resource_write_post             audit_pep_exec_microservice_pre
audit_pep_exec_rule_post                  audit_pep_resource_write_pre              audit_pep_resource_lseek_pre
audit_pep_resource_open_pre               audit_pep_resource_write_pre              audit_pep_exec_microservice_post
audit_pep_resource_lseek_post             audit_pep_resource_write_post             audit_pep_resource_lseek_post
audit_pep_resource_open_post              audit_pep_resource_write_pre              audit_pep_exec_rule_post
audit_pep_exec_rule_pre                   audit_pep_resource_write_post             audit_pep_resource_open_pre
    (acPreProcForServerPortal)            audit_pep_resource_open_post              audit_pep_resource_write_pre
audit_pep_exec_microservice_pre           audit_pep_exec_rule_pre                   audit_pep_resource_write_post
audit_pep_resource_lseek_pre                  (acPreProcForServerPortal)            audit_pep_resource_write_pre
audit_pep_exec_microservice_post          audit_pep_resource_lseek_pre              audit_pep_resource_write_post
audit_pep_resource_lseek_post             audit_pep_exec_microservice_pre           audit_pep_resource_write_pre
audit_pep_exec_rule_post                  audit_pep_exec_microservice_post          audit_pep_resource_write_pre
audit_pep_resource_open_pre               audit_pep_resource_lseek_post             audit_pep_resource_write_post
audit_pep_resource_open_post              audit_pep_exec_rule_post                  audit_pep_resource_write_pre
audit_pep_exec_rule_pre                   audit_pep_resource_open_pre               audit_pep_resource_write_post
    (acPreProcForServerPortal)            audit_pep_resource_write_pre              audit_pep_resource_write_pre
audit_pep_resource_lseek_pre              audit_pep_resource_write_post             audit_pep_resource_write_post
audit_pep_exec_microservice_pre           audit_pep_resource_write_pre              audit_pep_resource_write_pre
audit_pep_exec_microservice_post          audit_pep_resource_write_post             audit_pep_resource_write_post
audit_pep_resource_lseek_post             audit_pep_resource_write_pre              audit_pep_resource_write_pre
audit_pep_exec_rule_post                  audit_pep_resource_write_post             audit_pep_resource_write_post
audit_pep_resource_open_pre               audit_pep_resource_write_pre              audit_pep_resource_write_pre
audit_pep_resource_open_post              audit_pep_resource_write_post             audit_pep_resource_write_post
audit_pep_exec_rule_pre                   audit_pep_resource_write_pre              audit_pep_resource_write_pre
    (acPreProcForServerPortal)            audit_pep_resource_open_post              audit_pep_resource_write_post
audit_pep_resource_write_pre              audit_pep_resource_write_post             audit_pep_resource_write_pre
audit_pep_resource_lseek_pre              audit_pep_exec_rule_pre                   audit_pep_resource_write_post
audit_pep_exec_microservice_pre               (acPreProcForServerPortal)            audit_pep_resource_write_pre
audit_pep_resource_lseek_post             audit_pep_resource_lseek_pre              audit_pep_resource_write_post
audit_pep_resource_write_pre              audit_pep_exec_microservice_pre           audit_pep_resource_write_pre
audit_pep_exec_microservice_post          audit_pep_exec_microservice_post          audit_pep_resource_open_post
audit_pep_exec_rule_post                  audit_pep_resource_lseek_post             audit_pep_resource_write_post
audit_pep_resource_open_pre               audit_pep_exec_rule_post                  audit_pep_exec_rule_pre
audit_pep_resource_write_pre              audit_pep_resource_open_pre                   (acPreProcForServerPortal)
audit_pep_resource_write_post             audit_pep_resource_write_pre              audit_pep_resource_lseek_pre
audit_pep_resource_write_pre              audit_pep_resource_write_post             audit_pep_exec_microservice_pre
audit_pep_resource_write_pre              audit_pep_resource_write_pre              audit_pep_exec_microservice_post
audit_pep_resource_open_post              audit_pep_resource_write_pre              audit_pep_resource_lseek_post
audit_pep_resource_write_post             audit_pep_resource_write_post             audit_pep_exec_rule_post
audit_pep_exec_rule_pre                   audit_pep_resource_write_pre              audit_pep_resource_open_pre
    (acPreProcForServerPortal)            audit_pep_resource_write_post             audit_pep_resource_write_post
audit_pep_resource_lseek_pre              audit_pep_resource_write_pre              audit_pep_resource_write_pre
audit_pep_exec_microservice_pre           audit_pep_resource_write_pre              audit_pep_resource_write_pre
audit_pep_exec_microservice_post          audit_pep_resource_write_post             audit_pep_resource_write_post
audit_pep_resource_lseek_post             audit_pep_resource_write_pre              audit_pep_resource_write_pre
audit_pep_exec_rule_post                  audit_pep_resource_write_post             audit_pep_resource_write_pre
audit_pep_resource_open_pre               audit_pep_resource_write_pre              audit_pep_resource_write_post
audit_pep_resource_write_pre              audit_pep_resource_write_post             audit_pep_resource_write_pre
audit_pep_resource_write_post             audit_pep_resource_write_pre              audit_pep_resource_write_post
audit_pep_resource_write_pre              audit_pep_resource_write_post             audit_pep_resource_write_pre
audit_pep_resource_write_pre              audit_pep_resource_open_post              audit_pep_resource_write_post
audit_pep_resource_write_post             audit_pep_exec_rule_pre                   audit_pep_resource_write_pre
audit_pep_resource_write_pre                  (acPreProcForServerPortal)            audit_pep_resource_write_post
audit_pep_resource_write_post             audit_pep_resource_lseek_pre              audit_pep_resource_write_pre
audit_pep_resource_open_post              audit_pep_exec_microservice_pre           audit_pep_resource_write_post
audit_pep_exec_rule_pre                   audit_pep_exec_microservice_post          audit_pep_resource_write_pre
    (acPreProcForServerPortal)            audit_pep_resource_lseek_post             audit_pep_resource_write_post
audit_pep_exec_microservice_pre           audit_pep_exec_rule_post                  audit_pep_resource_write_pre
audit_pep_resource_lseek_pre              audit_pep_resource_open_pre               audit_pep_resource_write_post
audit_pep_exec_microservice_post          audit_pep_resource_write_pre              audit_pep_resource_write_pre
audit_pep_exec_rule_post                  audit_pep_resource_write_post             audit_pep_resource_write_post
audit_pep_resource_lseek_post             audit_pep_resource_write_pre              audit_pep_resource_write_pre
                                          audit_pep_resource_write_pre              audit_pep_resource_write_post
                                          audit_pep_resource_write_post             audit_pep_resource_write_pre
                                          audit_pep_resource_write_pre              audit_pep_resource_write_post
                                          audit_pep_resource_write_post             audit_pep_resource_open_post
                                          audit_pep_resource_write_pre              audit_pep_exec_rule_pre
                                          audit_pep_resource_write_post                 (acPreProcForServerPortal)
                                          audit_pep_resource_write_pre              audit_pep_resource_lseek_pre
                                          audit_pep_resource_write_post             audit_pep_exec_microservice_pre
                                          audit_pep_resource_open_post              audit_pep_exec_microservice_post
                                          audit_pep_exec_rule_pre
                                              (acPreProcForServerPortal)
                                          audit_pep_resource_lseek_pre
                                          audit_pep_exec_microservice_pre
                                          audit_pep_exec_microservice_post
                                          audit_pep_resource_lseek_post
                                          audit_pep_exec_rule_post
```

```
audit_pep_resource_lseek_post
audit_pep_exec_rule_post
audit_pep_resource_open_pre
audit_pep_resource_write_pre
audit_pep_resource_write_post
audit_pep_resource_write_pre
audit_pep_resource_write_pre
audit_pep_resource_write_post
audit_pep_resource_write_pre
audit_pep_resource_write_post
audit_pep_resource_write_pre
audit_pep_resource_write_post
audit_pep_resource_write_pre
audit_pep_resource_write_post
audit_pep_resource_write_pre
audit_pep_resource_write_post
audit_pep_resource_write_pre
audit_pep_resource_write_post
audit_pep_resource_write_pre
audit_pep_resource_write_post
audit_pep_resource_write_pre
audit_pep_resource_write_post
audit_pep_resource_write_pre
audit_pep_resource_write_post
audit_pep_resource_write_post
audit_pep_resource_write_pre
audit_pep_resource_write_pre
audit_pep_resource_open_post
audit_pep_resource_write_post
audit_pep_exec_rule_pre
    (acPreProcForServerPortal)
audit_pep_exec_microservice_pre
audit_pep_exec_microservice_post
audit_pep_resource_lseek_pre
audit_pep_exec_rule_post
audit_pep_resource_lseek_post
audit_pep_resource_open_pre
audit_pep_resource_write_pre
audit_pep_resource_write_post
audit_pep_resource_write_pre
audit_pep_resource_write_pre
audit_pep_resource_write_post
audit_pep_resource_write_pre
audit_pep_resource_write_post
audit_pep_resource_write_pre
audit_pep_resource_write_post
audit_pep_resource_write_pre
audit_pep_resource_write_post
audit_pep_resource_write_pre
audit_pep_resource_write_post
audit_pep_resource_write_pre
audit_pep_resource_write_post
audit_pep_resource_write_pre
audit_pep_resource_write_post
audit_pep_resource_write_pre
audit_pep_resource_write_post
audit_pep_resource_write_pre
audit_pep_resource_write_post
audit_pep_resource_open_post
audit_pep_resource_lseek_pre
audit_pep_resource_lseek_post
audit_pep_resource_write_pre
audit_pep_resource_write_post
audit_pep_resource_write_pre
audit_pep_resource_write_pre
audit_pep_resource_write_pre
audit_pep_resource_write_post
audit_pep_resource_write_pre
audit_pep_resource_write_post
audit_pep_resource_write_pre
audit_pep_resource_write_post
audit_pep_resource_write_post
audit_pep_resource_write_pre
audit_pep_resource_write_post
audit_pep_resource_write_pre
audit_pep_resource_write_post
audit_pep_resource_write_pre
audit_pep_resource_write_post
audit_pep_resource_write_pre
audit_pep_resource_write_post
audit_pep_resource_write_pre

audit_pep_resource_write_post
audit_pep_resource_write_post
audit_pep_resource_write_pre
audit_pep_resource_write_post
audit_pep_resource_write_pre
audit_pep_resource_write_post
audit_pep_resource_write_pre
audit_pep_resource_write_post
audit_pep_resource_write_pre
audit_pep_resource_write_post
audit_pep_resource_write_pre
audit_pep_resource_write_post
audit_pep_resource_write_pre
audit_pep_resource_write_post
audit_pep_resource_write_pre
audit_pep_resource_write_post
audit_pep_resource_write_pre
audit_pep_resource_write_post
audit_pep_resource_write_pre
audit_pep_resource_write_post
audit_pep_resource_write_pre
audit_pep_resource_write_post
audit_pep_resource_write_pre
audit_pep_resource_write_post
audit_pep_resource_write_pre
audit_pep_resource_write_post
audit_pep_resource_write_pre
audit_pep_resource_write_post
audit_pep_resource_write_pre
audit_pep_resource_write_post
audit_pep_resource_write_pre
audit_pep_resource_write_post
audit_pep_resource_write_pre
audit_pep_resource_write_post
audit_pep_resource_write_pre
audit_pep_resource_write_post
audit_pep_resource_write_pre
audit_pep_resource_write_post
audit_pep_resource_write_pre
audit_pep_resource_write_post
audit_pep_resource_write_pre
audit_pep_resource_write_post
audit_pep_resource_write_pre
audit_pep_resource_write_post
audit_pep_resource_write_pre
audit_pep_resource_write_post
audit_pep_resource_write_pre
audit_pep_resource_write_post
audit_pep_resource_write_pre
audit_pep_resource_write_post
audit_pep_resource_write_pre
audit_pep_resource_write_post
audit_pep_resource_write_pre
audit_pep_resource_write_post
audit_pep_resource_write_pre
audit_pep_resource_write_post
audit_pep_resource_write_post
audit_pep_resource_write_pre
audit_pep_resource_write_post
audit_pep_resource_write_pre
audit_pep_resource_write_pre
audit_pep_resource_write_post
audit_pep_resource_write_pre
audit_pep_resource_write_post
audit_pep_resource_write_pre
audit_pep_resource_write_post
audit_pep_resource_write_pre
audit_pep_resource_write_post
audit_pep_resource_write_pre
audit_pep_resource_write_post
audit_pep_resource_write_pre
audit_pep_resource_write_post
audit_pep_resource_write_pre
audit_pep_resource_write_pre
audit_pep_resource_write_pre
audit_pep_resource_write_post
audit_pep_resource_write_pre
audit_pep_resource_write_pre
audit_pep_resource_write_post
audit_pep_resource_write_pre

audit_pep_resource_write_post
audit_pep_resource_write_pre
audit_pep_resource_write_post
audit_pep_resource_write_pre
audit_pep_resource_write_post
audit_pep_resource_write_pre
audit_pep_resource_write_post
audit_pep_resource_write_pre
audit_pep_resource_write_post
audit_pep_resource_write_pre
audit_pep_resource_write_post
audit_pep_resource_write_pre
audit_pep_resource_write_post
audit_pep_resource_write_pre
audit_pep_resource_write_post
audit_pep_resource_write_pre
audit_pep_resource_write_post
audit_pep_resource_write_pre
audit_pep_resource_write_post
audit_pep_resource_write_pre
audit_pep_resource_write_post
audit_pep_resource_write_pre
audit_pep_resource_write_post
audit_pep_resource_write_pre
audit_pep_resource_write_post
audit_pep_resource_write_pre
audit_pep_resource_write_post
audit_pep_resource_write_pre
audit_pep_resource_write_post
audit_pep_resource_write_pre
audit_pep_resource_write_post
audit_pep_resource_write_pre
audit_pep_resource_write_post
audit_pep_resource_write_pre
audit_pep_resource_write_post
audit_pep_resource_write_pre
audit_pep_resource_write_post
audit_pep_resource_write_pre
audit_pep_resource_write_post
audit_pep_resource_write_pre
audit_pep_exec_rule_pre
    (acPostProcForServerPortal)
audit_pep_exec_microservice_pre
audit_pep_exec_microservice_post
audit_pep_exec_rule_post
audit_pep_resource_close_pre
audit_pep_resource_close_post
audit_pep_resource_write_post
audit_pep_resource_write_post
audit_pep_resource_write_pre
audit_pep_resource_write_pre
audit_pep_resource_write_post
audit_pep_resource_write_pre
audit_pep_resource_write_post
audit_pep_resource_write_pre
audit_pep_resource_write_post
audit_pep_resource_write_pre
audit_pep_resource_write_post
audit_pep_resource_write_pre
audit_pep_resource_write_post
audit_pep_resource_write_pre
audit_pep_resource_write_post
audit_pep_resource_write_post
audit_pep_resource_write_pre
audit_pep_resource_write_pre
audit_pep_resource_write_post
audit_pep_resource_write_post
audit_pep_exec_rule_pre
    (acPostProcForServerPortal)
audit_pep_exec_microservice_pre
audit_pep_exec_microservice_post
audit_pep_exec_rule_post
audit_pep_resource_close_pre
audit_pep_resource_close_post
audit_pep_resource_write_post
audit_pep_resource_write_pre
audit_pep_resource_write_post
audit_pep_resource_write_pre
audit_pep_resource_write_post
audit_pep_resource_write_pre
```

audit_pep_resource_write_post
audit_pep_resource_write_pre
audit_pep_resource_write_post
audit_pep_exec_rule_pre
    (acPostProcForServerPortal)
audit_pep_exec_microservice_pre
audit_pep_exec_microservice_post
audit_pep_exec_rule_post
audit_pep_resource_close_pre
audit_pep_resource_close_post
audit_pep_resource_write_post
audit_pep_resource_write_pre
audit_pep_resource_write_post
audit_pep_resource_write_pre
audit_pep_resource_write_post
audit_pep_resource_write_pre
audit_pep_resource_write_post
audit_pep_resource_write_pre
audit_pep_resource_write_post
audit_pep_exec_rule_pre
    (acPostProcForServerPortal)
audit_pep_exec_microservice_pre
audit_pep_exec_microservice_post
audit_pep_exec_rule_post
audit_pep_resource_close_pre
audit_pep_resource_close_post
audit_pep_resource_write_post
audit_pep_resource_write_pre
audit_pep_resource_write_post
audit_pep_resource_write_pre
audit_pep_resource_write_post
audit_pep_resource_write_pre
audit_pep_resource_write_post
audit_pep_resource_write_pre
audit_pep_resource_write_post
audit_pep_exec_rule_pre
    (acPostProcForServerPortal)
audit_pep_exec_microservice_pre
audit_pep_exec_microservice_post
audit_pep_exec_rule_post
audit_pep_resource_close_pre
audit_pep_resource_close_post
audit_pep_resource_write_post
audit_pep_resource_write_pre
audit_pep_resource_write_post
audit_pep_resource_write_pre
audit_pep_resource_write_post
audit_pep_resource_write_pre
audit_pep_resource_write_post
audit_pep_exec_rule_pre
    (acPostProcForServerPortal)
audit_pep_exec_microservice_pre
audit_pep_exec_microservice_post
audit_pep_exec_rule_post
audit_pep_resource_close_pre
audit_pep_resource_close_post
audit_pep_resource_write_post
audit_pep_resource_write_pre
audit_pep_resource_write_post
audit_pep_resource_write_pre
audit_pep_resource_write_post
audit_pep_resource_write_pre
audit_pep_resource_write_post
audit_pep_resource_write_pre
audit_pep_resource_write_pre
audit_pep_resource_write_post
audit_pep_resource_write_pre
audit_pep_resource_write_post
audit_pep_resource_write_pre

audit_pep_resource_write_post
audit_pep_resource_write_pre
audit_pep_resource_write_post
audit_pep_resource_write_pre
audit_pep_resource_write_post
audit_pep_resource_write_pre
audit_pep_resource_write_post
audit_pep_resource_write_post
audit_pep_resource_write_pre
audit_pep_resource_write_post
audit_pep_resource_write_pre
audit_pep_resource_write_post
audit_pep_resource_write_pre
audit_pep_resource_write_post
audit_pep_resource_write_pre
audit_pep_resource_write_post
audit_pep_resource_write_pre
audit_pep_resource_write_post
audit_pep_resource_write_pre
audit_pep_resource_write_post
audit_pep_resource_write_pre
audit_pep_resource_write_post
audit_pep_resource_write_pre
audit_pep_resource_write_post
audit_pep_resource_write_pre
audit_pep_resource_write_post
audit_pep_resource_write_pre
audit_pep_resource_write_post
audit_pep_exec_rule_pre
    (acPostProcForServerPortal)
audit_pep_exec_microservice_pre
audit_pep_exec_microservice_post
audit_pep_exec_rule_post
audit_pep_resource_close_pre
audit_pep_resource_close_post
audit_pep_resource_write_post
audit_pep_exec_rule_pre
    (acPostProcForServerPortal)
audit_pep_exec_microservice_pre
audit_pep_exec_microservice_post
audit_pep_exec_rule_post
audit_pep_resource_close_pre
audit_pep_resource_close_post
audit_pep_resource_write_post
audit_pep_resource_write_pre
audit_pep_resource_write_post
audit_pep_resource_write_pre
audit_pep_resource_write_post
audit_pep_resource_write_pre
audit_pep_resource_write_post
audit_pep_resource_write_pre
audit_pep_resource_write_post
audit_pep_resource_write_pre
audit_pep_resource_write_post
audit_pep_resource_write_pre
audit_pep_resource_write_post
audit_pep_exec_rule_pre
    (acPostProcForServerPortal)
audit_pep_exec_microservice_pre
audit_pep_exec_microservice_post
audit_pep_exec_rule_post
audit_pep_resource_close_pre
audit_pep_resource_close_post
audit_pep_resource_write_post
audit_pep_resource_write_pre
audit_pep_resource_write_post
audit_pep_resource_write_pre
audit_pep_resource_write_post
audit_pep_resource_write_pre
audit_pep_resource_write_post
audit_pep_resource_write_pre
audit_pep_resource_write_post
audit_pep_resource_write_post
audit_pep_resource_write_pre
audit_pep_resource_write_post
audit_pep_resource_write_pre
audit_pep_resource_write_post
audit_pep_resource_write_pre
audit_pep_resource_write_post
audit_pep_resource_write_post
audit_pep_resource_write_post
audit_pep_resource_write_pre
audit_pep_resource_write_post
audit_pep_resource_write_pre
audit_pep_resource_write_post
audit_pep_exec_rule_pre

    (acPostProcForServerPortal)
audit_pep_exec_microservice_pre
audit_pep_exec_microservice_post
audit_pep_exec_rule_post
audit_pep_resource_close_pre
audit_pep_resource_close_post
audit_pep_resource_write_post
audit_pep_resource_write_pre
audit_pep_resource_write_post
audit_pep_resource_write_pre
audit_pep_resource_write_post
audit_pep_resource_write_pre
audit_pep_resource_write_post
audit_pep_resource_write_pre
audit_pep_resource_write_post
audit_pep_resource_write_pre
audit_pep_resource_write_post
audit_pep_resource_write_pre
audit_pep_resource_write_post
audit_pep_resource_write_pre
audit_pep_resource_write_post
audit_pep_exec_rule_pre
    (acPostProcForServerPortal)
audit_pep_exec_microservice_pre
audit_pep_exec_microservice_post
audit_pep_exec_rule_post
audit_pep_resource_close_pre
audit_pep_resource_close_post
audit_pep_resource_write_post
audit_pep_resource_write_pre
audit_pep_resource_write_post
audit_pep_resource_write_pre
audit_pep_resource_write_post
audit_pep_exec_rule_pre
    (acPostProcForServerPortal)
audit_pep_exec_microservice_pre
audit_pep_exec_microservice_post
audit_pep_exec_rule_post
audit_pep_resource_close_pre
audit_pep_resource_close_post
audit_pep_resource_write_post
audit_pep_resource_write_pre
audit_pep_resource_write_post
audit_pep_resource_write_pre
audit_pep_resource_write_post
audit_pep_resource_write_pre
audit_pep_resource_write_post
audit_pep_resource_write_pre
audit_pep_resource_write_post
audit_pep_resource_write_pre
audit_pep_resource_write_post
audit_pep_resource_write_pre
audit_pep_resource_write_post
audit_pep_resource_write_pre
audit_pep_resource_write_post
audit_pep_resource_write_pre
audit_pep_resource_write_post
audit_pep_resource_write_pre
audit_pep_resource_write_post
audit_pep_exec_rule_pre
    (acPostProcForServerPortal)
audit_pep_exec_microservice_pre
audit_pep_exec_microservice_post
audit_pep_exec_rule_post
audit_pep_resource_close_pre
audit_pep_resource_close_post
audit_pep_resource_write_post
audit_pep_resource_write_pre
audit_pep_resource_write_post
audit_pep_resource_write_pre
audit_pep_resource_write_post
audit_pep_resource_write_pre
audit_pep_resource_write_post
audit_pep_resource_write_pre
audit_pep_resource_write_post
audit_pep_resource_write_pre
audit_pep_resource_write_post
audit_pep_resource_write_pre
audit_pep_resource_write_post
audit_pep_resource_write_pre
audit_pep_resource_write_post
audit_pep_resource_write_pre
audit_pep_resource_write_post
audit_pep_resource_write_pre
audit_pep_resource_write_post
audit_pep_exec_rule_pre
    (acPostProcForServerPortal)
audit_pep_exec_microservice_pre

```
audit_pep_exec_microservice_post
audit_pep_exec_rule_post
audit_pep_resource_close_pre
audit_pep_resource_close_post
audit_pep_resource_write_post
audit_pep_resource_write_pre
audit_pep_resource_write_post
audit_pep_exec_rule_pre
       (acPostProcForServerPortal)
audit_pep_exec_microservice_pre
audit_pep_exec_microservice_post
audit_pep_exec_rule_post
audit_pep_network_read_header_pre
audit_pep_network_read_header_post
audit_pep_network_read_body_pre
audit_pep_network_read_body_post
audit_pep_opr_complete_pre
audit_pep_resource_close_pre
audit_pep_resource_close_post
audit_pep_resource_stat_pre
audit_pep_resource_stat_post
audit_pep_exec_rule_pre
       (acPreProcForModifyDataObjMeta)
audit_pep_exec_microservice_pre
audit_pep_exec_microservice_post
audit_pep_exec_rule_post
audit_pep_database_mod_data_obj_meta_pre
audit_pep_database_mod_data_obj_meta_post
audit_pep_exec_rule_pre
       (acPostProcForModifyDataObjMeta)
audit_pep_exec_microservice_pre
audit_pep_exec_microservice_post
audit_pep_exec_rule_post
audit_pep_resource_modified_pre
audit_pep_resource_modified_post
audit_pep_exec_rule_pre
       (acPostProcForPut)
audit_pep_exec_microservice_pre
audit_pep_exec_microservice_post
audit_pep_exec_rule_post
audit_pep_opr_complete_post
audit_pep_network_write_body_pre
audit_pep_network_write_header_pre
audit_pep_network_write_header_post
audit_pep_network_write_body_post
audit_pep_data_obj_put_post
audit_pep_auth_agent_start_pre
audit_pep_auth_agent_start_post
audit_pep_network_read_header_pre
audit_pep_network_read_header_post
audit_pep_network_read_body_pre
audit_pep_network_read_body_post
audit_pep_network_agent_stop_pre
audit_pep_network_agent_stop_post
audit_pep_database_close_pre
audit_pep_database_close_post
```

VISUAL DIFFERENCE OF `ireg` AND `iput`

Once a full audit of the iCommands are readily available, it becomes clear that cursory understandings of internal iRODS functionality can be interrogated. One of these unexamined understandings may have been the relationship of `ireg` with `iput`. Figure 1 compares the two and shows that an `ireg` is a subset of an `iput` (permissions are checked and the data is transferred before the registration into the catalog).

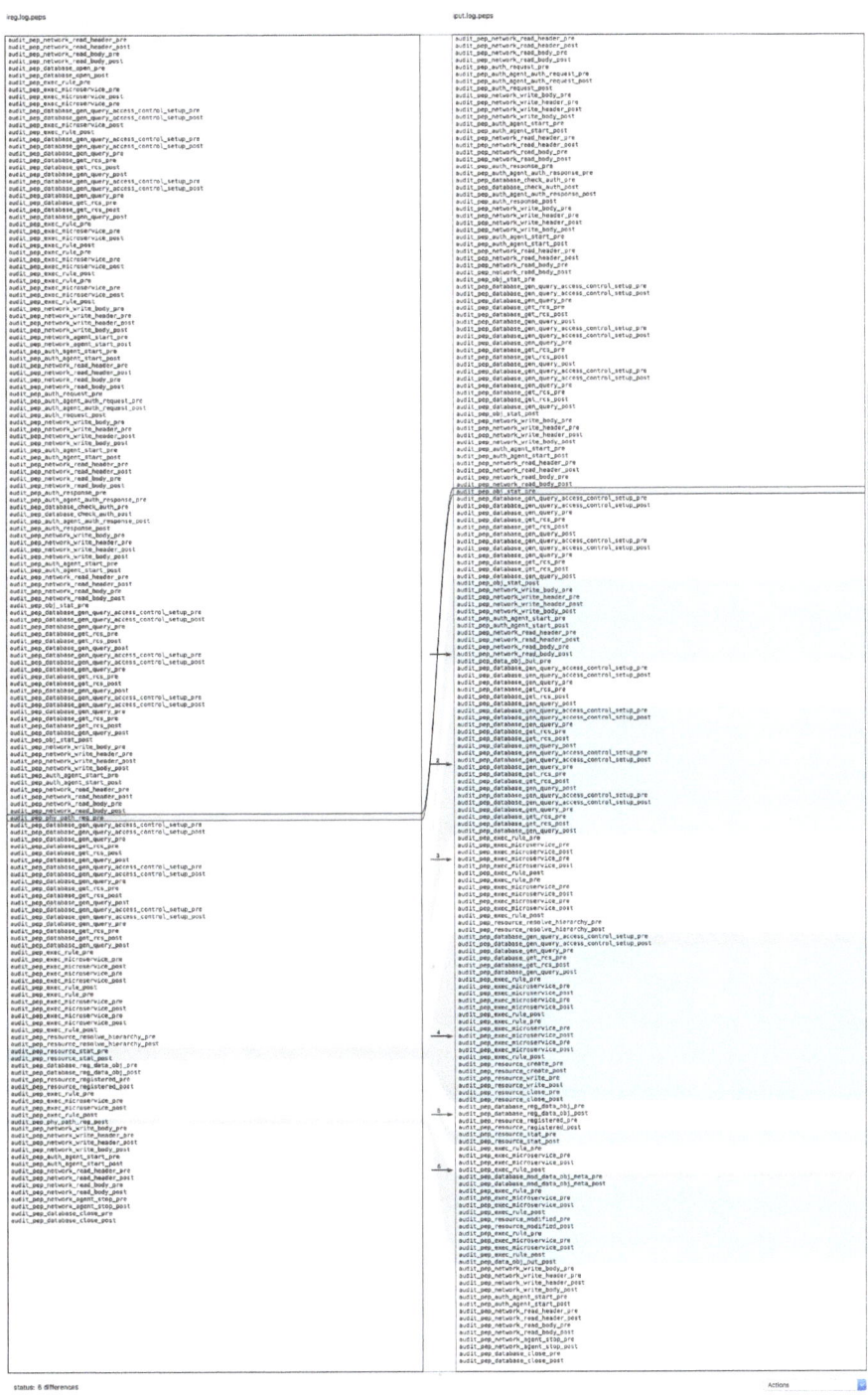

Figure 1. The `ireg` PEPs are a subset of the `iput` PEPs

AGGREGATE DISPLAY OF AMQP MESSAGES

Since every client connection and every operation within the server can be audited, it is useful to visualize and quantify this activity. The following initial dashboard data was generated by a small script that moved files around in an iRODS 4.2 Zone consisting of two servers.

The pipeline (Figure 2) consists of a connection between iRODS and ActiveMQ (or Apollo) and then configuring the Elastic Stack (Logstash, Elasticsearch, and Kibana) to pull the messages and chart them.

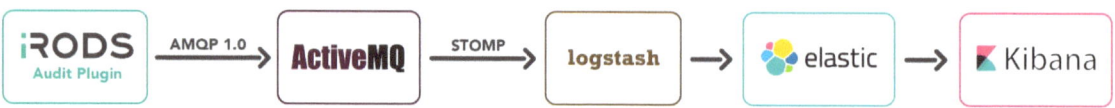

Figure 2. AMQP data pipeline from iRODS to Kibana

Shown in Figure 3 are Connections, Unique Users, Bytes Written, Bytes Read (all per minute, and per server). Additionally, aggregate Top Client IPs and Top Users are displayed across the entire Zone.

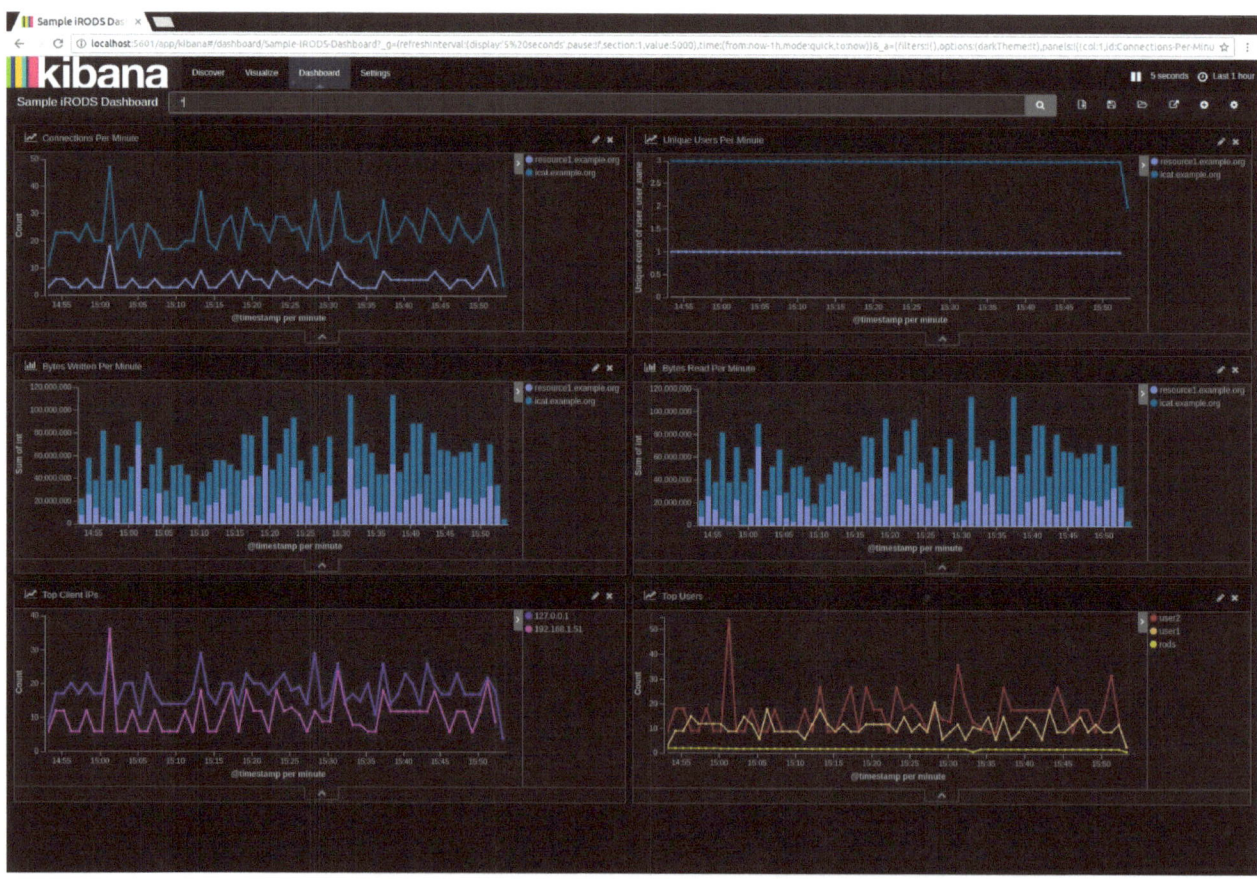

Figure 3. Initial Kibana dashboard using data from the iRODS Audit AMQP rule engine plugin

CONCLUSION

iRODS 4.2 represents a significant effort towards providing more integration-friendly functionality for distributed policy-based data management. iRODS 4.2 can now be a part of enterprise-quality metrics gathering and reporting.

This example has shown only a basic AMQP message bus, but with the pluggable rule engine, multiple plugin instances can be run concurrently and emit onto targeted topics or queues. This allows for different queries and assumptions to be tested on live data without getting in the way of the iRODS servers themselves.

By default, the audit plugin is configured to use a queue (store-and-forward) to guarantee that no messages are lost, but it can easily be configured to use a topic or topics which would allow for a broadcast style of messaging if multiple servers are needed to do different analyses on the same messages.

The AMQP messages themselves can be directed to other types of reporting or indexing technology as well. A greater goal of iRODS 4.2 is to allow external services to react to different types of activity within an iRODS deployment. When a file is created or updated, a message could be sent to an indexing server which could then retrieve the full text of the data object within iRODS and produce and maintain a full-text index.

The AMQP audit plugin is a powerful new addition to the iRODS technology platform and should dramatically aid in the automation and instrumentation of complex data grids.

Speed Research Discovery
with Comprehensive Storage and Data Management

Highlights

Comprehensive technical computing storage and data management architecture

HGST Active Archive System

- Simple to Deploy – Power and network connections are all you need

- Extreme Scale – Increase capacity and performance in line with data growth

- Highest Resiliency – up to 15 nines data durability, with the ability to survive a data center outage in in 3-geo-configuration

- Enterprise Security – end-to-end encryption security for in-flight and data at-rest protection

- Excellent TCO – Low acquisition cost, power/TB, high capacity and density

Panasas ActiveStor

- Lightning-fast response time and parallel access for massive throughput

- Scales to 12PB and 150GB/s or 1.4M IOPS

iRODS Data Management Software

- Rules-based open source software

- Workflow automation with rules engine

- Easy data discovery with metadata catalog

- Rules-based storage tiering for efficiency

The rate of progress in life sciences research is accelerating exponentially leading to important advances in healthcare, agriculture, climate science and more, but those advances also create a mountain of data. IT managers supporting these efforts are being challenged to provide researchers the right solution that will accelerate their work. Complex simulations can take days or weeks to run. When simulations take longer, discoveries are delayed slowing analysis and ultimately the commercialization of the findings. Faster simulations allow more complex models to be run – leading to improved discoveries through bioinformatics.

A key to accelerating life science application performance is implementing a high performance computing (HPC) infrastructure that eliminates computing and storage bottlenecks, enables better collaboration, and preserves simplicity so researchers focus their efforts on discovery.

Conventional Storage is Slowing the Progress

Storage performance and data management are major causes of computing bottlenecks. The volume and type of data generated by modern lab equipment along with varying application and workflow requirements, makes implementing the right solution all the more challenging. In some cases, data is generated in one place and kept there, while other times the data is generated by many researchers around the world whose results and expertise must be pooled together to achieve the biggest benefit. Furthermore, HPC environments place special demands on storage with compute clusters that can have hundreds of nodes and thousands of cores working in parallel. Technical computing applications tend to be I/O bound with large numbers of sequential and random rapid read/write operations that can exhaust conventional storage, resulting in workflow bottlenecks, costly islands of storage, increased management effort, and longer time-to-discovery. A better approach is needed.

Eliminate Storage Performance Bottlenecks with Parallel Access

Panasas has an advanced performance scale-out NAS solution called ActiveStor that is designed to maximize mixed workload performance in HPC environments. Based on a fifth-generation architecture and parallel file system, application clients have simultaneous fast direct parallel access to large and growing datasets, avoiding the need to copy datasets locally to the compute cluster prior to processing. Direct parallel data access is also important because in the case of genomics, while sequencers often generate data in single streams, analysis of sequencer data can be done in parallel with many clients reading and writing directly to storage. In addition, up to 90% of metadata-related operations happen outside the data path, minimizing data access impact, resulting in faster workflows.

Object Storage for Massive Data Growth and Global Collaboration

Built using next generation object storage technology, the HGST Active Archive System enables research organizations to help cost-effectively manage enormous data growth. Serving as the capacity optimized secondary archive tier (Figure 1), the system's industry leading durability and data integrity makes it ideal for long-term data preservation. The system is simple to deploy and manage and can be easily scaled over time. IT management overhead is minimized with automated self-healing and proactive data integrity checks. Deployed in 3-site-geospread configuration, data is efficiently spread across three sites making it ideal for collaboration across distributed researchers. Integration with HPC workflows is easy using iRODS and its S3 resource server plugin.

Data Management at Scale in a Distributed Environment

Data-intensive technical computing applications such as in life sciences, require efficient, secure and cost effective data management in a wide-area distributed environment. Datasets are often shared by a global community of researchers that need to easily find and transfer subsets of data to a local or remote resource such as a private or public cloud for further processing. Open source data management software called iRODS (Integrated Rule-Oriented Data System) is increasingly used in a variety of technical computing workflows.

iRODS virtualizes data storage resources by putting a unified namespace on files regardless of where they are located, making them appear to the user as one project. Data discovery is made easy using a metadata catalog that describes every file, every directory, and every storage resource in the iRODS data grid, including federated grids (Figure 1). Data workflows can be automated with a rules engine that permits any action to be initiated by any trigger on any server or client in the grid. Collaboration is made secure and easy with configurable user authentication and SSL, users need only to be logged in to their home grid to access data on a remote grid. iRODs also helps control costs by allowing data to be aligned to the right storage technology through rules-based tiering. For instance, the majority of data can be stored on a capacity optimized object storage active archive tier and automatically moved to the high-performance scale-out file tier – significantly reducing CapEx.

Learn more about HGST Active Archive System at www.hgst.com/activearchive

Learn more about Panasas ActivStor at www.panasas.com

Learn more about iRODS software at www.irods.org

Life Sciences / HPC Storage and Data Managment Architecture

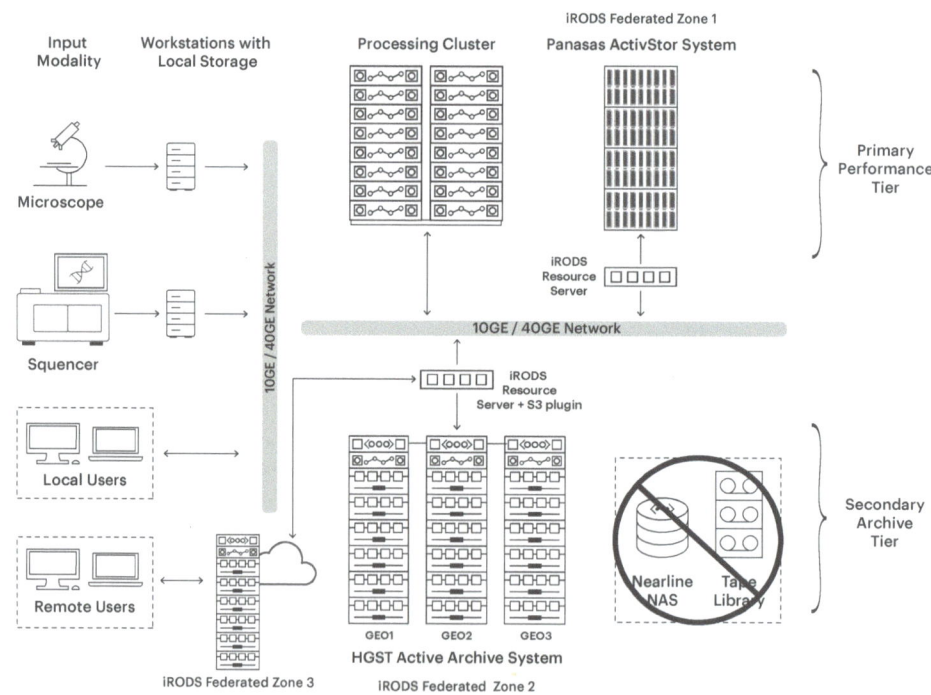

Figure 1. *Life sciences and HPC storage architecture example with iRODS federated data grids*

Integrating HUBzero and iRODS: Geospatial Data Management for Collaborative Scientific Research

Rajesh Kalyanam
Purdue University
rkalyana@purdue.edu

Robert A. Campbell
Purdue University
rcampbel@purdue.edu

Samuel P. Wilson
Purdue University
spwilson@purdue.edu

Pascal Meunier
Purdue University
pmeunier@purdue.edu

Lan Zhao
Purdue University
lanzhao@purdue.edu

Elizabett A. Hillery
Purdue University
eahillery@purdue.edu

Carol Song
Purdue University
carolxsong@purdue.edu

ABSTRACT

Geospatial data is now increasingly used with tools in diverse fields such as agronomy, hydrology and sociology to gain a better understanding of scientific data. Funded by the NSF DIBBS program, the GABBS project seeks to create reusable building blocks aiding researchers in adding geospatial data processing, visualization and curation to their tools. GABBS leverages the HUBzero cyberinfrastructure platform and iRODS to build a web-based collaborative research platform with enhanced geospatial capabilities. HUBzero is unique in its availability of a rapid tool development kit that simplifies web-enabling existing tools. Its support for dataset DOI association enables citable tool results. In short, it provides a seamless path from data collection, to simulation and publication and can benefit from iRODS data management at each step. Scientific tools often require and generate metadata with their outputs. Given the structured nature of geospatial data, automatic metadata capture is vital in avoiding repetitive work. iRODS microservices enable this automation of data processing, metadata capture and indexing for searchability. They also allow for similar offline ingestion of external research data. The iRODS FUSE filesystem mounts directly onto the hub, enabling tools to refer to local file paths, simplifying development. In this paper we discuss this work of integrating iRODS with HUBzero in the GABBS project and share our experience and lessons learned from this endeavor.

Keywords

Cyberinfrastructure, geospatial data, collaborative research.

INTRODUCTION

Researchers in the agronomy, hydrology and climate sciences fields are increasingly converging to collaborate on projects that leverage their combined expertise. In addition to originating from diverse fields, they are typically distributed globally and often have to share data and analysis tools with each other. For instance, the recent projects, DriNet [1], Geoshare [2], and U2U [3,4] have involved researchers from the agricultural, climate sciences, economics, hydrology fields and more, studying the effects of climate change on crop yields and drought prediction. Because of the distributed nature of collaborators, part of the focus in such projects is then justifiably placed on building a cyberinfrastructure platform enabling researchers to collaborate, contribute data and tools and support basic social media interaction. Rather than mainly being a simple data repository that users can upload to or download data from, these platforms need to support an intuitive interface for exploring the data, support a seamless integration of processing and visualization tools with the data and finally enable the curation of data to be used in

iRODS UGM 2016, June 8-9, 2016, Chapel Hill, NC.

scientific publications. Having all these capabilities on a single web-accessible platform saves collaborators from the effort of having to obtain and install these various tools on their local machines or handle data transfer between each of these steps. This end-to-end pipeline involving data collection, discovery, analysis, and publication can benefit from a robust data management system that provides several client libraries for ease of integration into the various components of a cyberinfrastructure platform.

Geospatial data often plays an important role in these collaborative efforts. Global satellite and climate data can aid in better predictions of crop yields, droughts and water quality. For instance, the percentage of snow coverage on the ground in a particular year affects the sowing schedule and subsequently, the number of days that the crop can grow before the next harvest. This is directly correlated to the crop yields. Research models are now being built to take these additional factors into account to improve their accuracy, making access to geospatial data vital to such analysis. However, the scientists collecting and processing the geospatial data are often different from the users (agronomists, for example) of such data. There is a need to make the relevant geospatial data easy to discover, visualize and transform. Visualizing geospatial data provides both a quick way to verify that it is from the region of interest, and an intuitive understanding of the correlation between data from disparate domains when viewed in conjunction with data such as crop yields. Thus, there is a need to augment the data storage with specialized processing that can support such operations on data.

The GABBS (Geospatial Analysis Building Blocks) project aims to support collaborative research involving geospatial data by creating the necessary reusable pieces that such efforts can utilize. In particular, this includes geospatial data management, automatic metadata extraction, geo-located search, geospatial data visualization tools and general-purpose map development APIs. This would make it possible for new collaborative efforts to use these building blocks rather than start from scratch. The GABBS building blocks are intended for installation on top of the open-source HUBzero cyberinfrastructure platform [5]. HUBzero, in addition to enabling user and community management, supports rapid tool development using its Rappture tool development kit. This makes it possible to quickly wrap a graphical user interface around scientific tools and deploy them onto a *hub* for use by other members of the hub. In the GABBS project, Rappture was extended to allow for map and geospatial visualization elements to be added to tools.

This paper describes our efforts towards building a collaborative research platform by integrating iRODS data management with HUBzero with special focus on geospatial data in support of the GABBS project goals.

HUBZERO PLATFORM

The HUBzero cyberinfrastructure platform grew out of efforts to create a general-purpose software platform that can be used to build powerful websites termed *hubs* that can be easily customized to suit the needs of a particular domain. The HUBzero Content Management System (CMS) enables users to create project groups, articles, blog entries and discussion groups. However, a primary focus of HUBzero has also been to enable collaborative scientific research. Users can contribute datasets and tools to the hub and create citable resources (with DOIs) that can be included in scientific publications. Such citable resources can range from datasets to scientific tools that can be run on the hub, leading to reproducible research. HUBzero was originally created at Purdue University in conjunction with the Network for Computational Nanotechnology (NCN) in support of nanoHUB [6], a platform for students and researchers alike to contribute, explore and use various modeling, visualization and simulation tools for nanotechnology. Faculty and researchers in several universities have successfully adopted nanoHUB for their nanotechnology research and education needs. HUBzero has grown in the past several years to support hubs for a diverse range of domains including, pharmaceuticals, cancer research, earthquake simulation, climate modeling and several others.

The primary collaboration space on a hub is a project where access is restricted to the members of that project. Projects provide a file management space where users can upload files, which are then available to all project members. Project files can then be selected to be included in publications with an associated DOI. However, project files provide limited processing and visualization capabilities. HUBzero by default provides a basic annotation interface, allowing metadata to be attached to files. Similarly, basic file types like images, PDFs and text files can be previewed to explore their contents.

One of the key factors enabling scientific research on a hub is the ability to create and deploy tools that can leverage high performance computing resources to submit simulation jobs. The Rappture tool development kit provides common GUI elements like dropdown lists, input boxes, plots and image viewers that can be composed together in a sandbox environment. The Rappture API (available in several programming languages) supports event handling on these elements in addition to input and output processing. Legacy scientific code can utilize this toolkit to add GUI elements that both simplify tool usage and add value via intuitive output visualizations. In addition to Rappture-based tools, scientific tools written in a variety of languages can be deployed onto a hub with minimal changes to ensure that the tool can run in the hub's execution environment. All hub tools are run in OpenVZ containers with VNC support and can be accessed through a web browser. This helps secure the user's tool session while restricting the resources available to the tool. These containers are designed to disallow arbitrary network connections and provide external data access via mounted file-systems. However, the primary input and output source for hub tools is typically the user's home directory on the hub. A file transfer utility is provided that enables file transfer between the user's home directory on the hub and their local desktop. This severely restricts the sizes of files that can be used with such tools. Users requiring more storage have to go through the process of requesting additional storage on their hub file space. Similarly, the project's file space is not directly linked to hub tools, requiring users to transfer files between their hub file space and their projects to enable sharing of tool results. Our work integrating iRODS with HUBzero is intended to overcome these shortcomings by providing an easily extensible, central storage system that can be accessed seamlessly in both the project space and hub tools.

THE GABBS PROJECT

The GABBS project was funded under the NSF DIBBS (Data Infrastructure Building Blocks) initiative that seeks to combine data-centric capabilities and services with cyberinfrastructure to foster collaborative scientific research. GABBS focuses on the task of simplifying the integration of geospatial data into domains that have not traditionally dealt with such data. The intent is to provide reusable (and easily accessible) blocks that support geospatial data processing, visualization, search and map-enabling locative data. As the project progresses, various GABBS infrastructure, components and tools are hosted on a production hub, i.e., MyGeoHub (http://mygeohub.org), and accessible publicly. Examples of GABBS-enabled tools include the MultiSpec and GeoBuilder tools.

The MultiSpec desktop tool that supports the visualization and processing of a large range of geospatial formats was deployed as a hub tool, allowing web access to a tool that would have otherwise required local installation. MultiSpec already had a few thousand users prior to being deployed onto MyGeoHub. It is expected that the simplified web-access will attract more users in the future. The GeoBuilder tool that serves as proof-of-concept of the new geospatial Rappture APIs was similarly deployed onto the hub. This tool demonstrates how a hub user can use the new Rappture map API and widgets to quickly construct a tool that provides a choice of base map layers and the ability to overlay locative tabular data on the base map. General-purpose options can be included for various map operations including, toggling zoom levels, filtering the displayed tabular data, selecting fields to be displayed and managing layer opacity. An open-source Python mapping library (PyMapLib) based on QGIS was also developed for tools requiring more complex geospatial capabilities, for instance, managing tiled layers, simple geospatial processing, value inspection and style customization. This mapping library is independent of HUBzero (unlike Rappture) and can thus have broader impact outside of the hub environment.

The central piece that connects all of these capabilities together is the ability to better manage geospatial data, making it seamlessly available to all the other GABBS components. This includes, automatic metadata extraction that can inform the user of useful details about the data, on-demand visualization that provides users with a quick overview of the data, data services enabling inter-operability with external applications and geospatial search that allows users to employ general-purpose keyword and bounding box searches to discover relevant data. The rest of this paper will describe how integrating iRODS with HUBzero allows us to provide these data capabilities while enabling a straightforward integration with hub tools.

IRODS IN A HUB

There are two data-intensive components of a hub that provide natural integration points for iRODS. First, the hub projects file space where users can contribute and share datasets and, second, the input source and output destination for hub tools. As mentioned previously, rather than just having a data repository supporting upload and download, our intent is to allow users to discover, explore, visualize and process the data. This involves extracting metadata enabling search, and converting data to a form that enables visualization using general-purpose mapping libraries. Locating these operations as close to the data source as possible improves the efficiency as well as distribution by simplifying packaging. Our integration of these capabilities into iRODS is described next. The overall system design is illustrated in Figure 1.

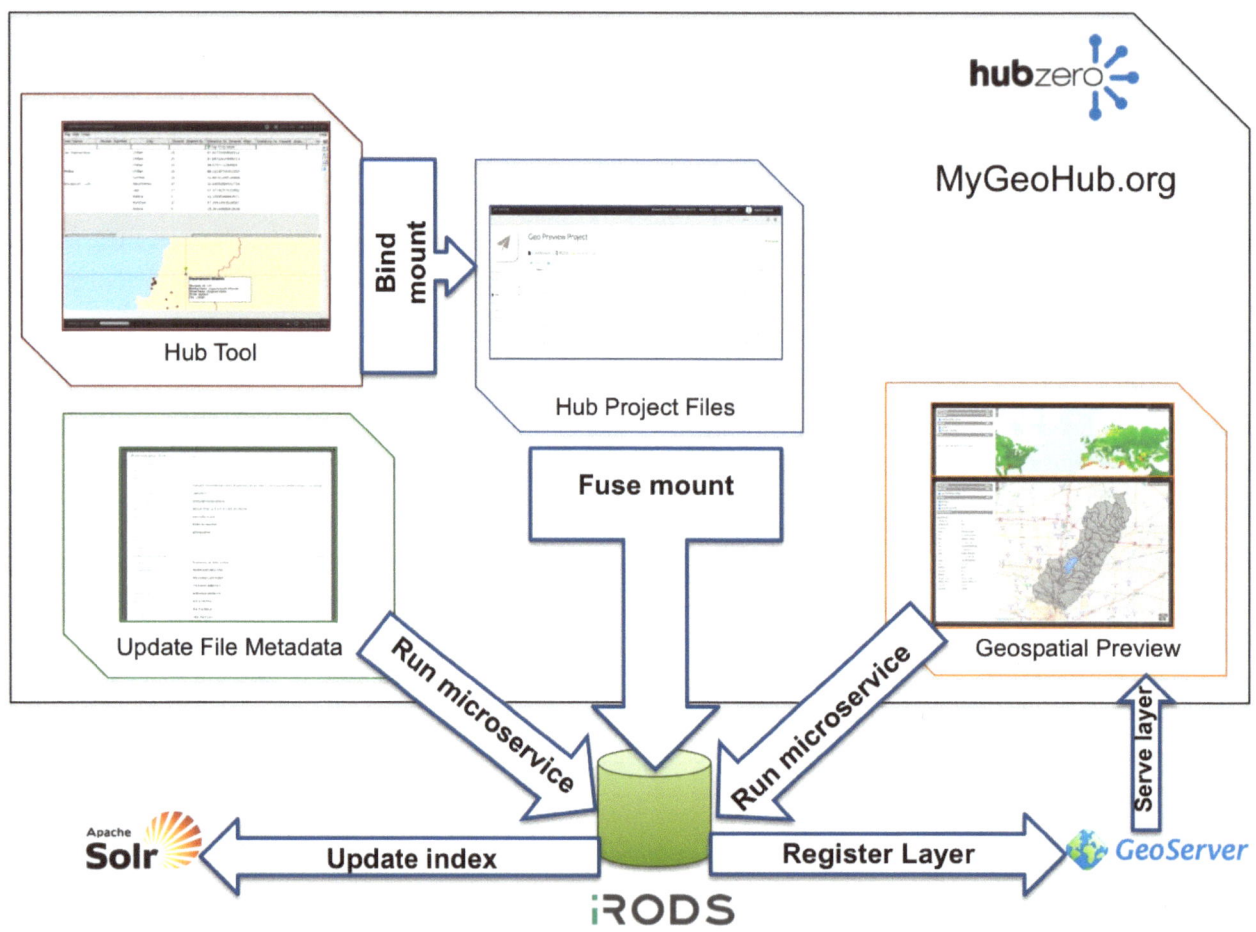

Figure 1. System Design

iRODS Filesystem Mounts

The hub projects file space is managed by the HUBzero CMS, written in PHP. One of the design considerations was potential future support for various file management software packages in this space, including Dropbox, Google Drive and Globus. HUBzero by default provides a local Git repository for each project for version-controlled storage. As a general-purpose solution, the PHP Flysystem adapter was employed to support these different storage systems. Rather than develop an iRODS Flysystem adapter from scratch, it was decided to use Flysystem's Local Filesystem adapter to manage iRODS collections mounted onto the hub webserver's filesystem. The iRODS FUSE client was used to mount iRODS collections owned by a single service account onto the hub webserver. Each hub project has its own top-level collection created when the project is created. Access control is then inherited from hub project membership. The hub project files user interface is designed to only allow access to files in that project's collection.

One of the shortcomings identified before is that datasets contributed to the project file space are not directly available to hub tools. To overcome this, the hub middleware code responsible for starting tool containers was modified to mount the projects file space in tool containers. Rather than separately FUSE mount the iRODS collections into each container, a *bind mount* is used to mount some part of the iRODS mount path to each tool container. In particular, only the folders corresponding to the projects that the current user is a member of are mounted into any tool session for that user. The project collections share the same name as the project, making such filtering straightforward. The availability of the project file space in a tool container session solves the issue of enabling project files as both input sources and output destinations for tools. More significantly, a tool developer does not need to code anything specific to leverage this; a tool can directly reference the local file path (in the container filesystem) where the bind mount has been created. This ability to maintain a single data repository linked to hub projects that can also function as the data repository for hub tools greatly simplifies data management in a hub, enabling a seamless link from data collection, sharing, processing and eventual publication.

Specialized Data Processing in iRODS

In order to make the most of the geospatial data managed in the hub, the iRODS storage needs to be augmented in support of metadata extraction, indexing for search and visualization. We accomplish this by attaching processing procedures via iRODS microservices. This has several advantages. First, the processing is located as close as possible to the data source, avoiding the need for data transfer. Second, these procedures can be packaged and easily installed via iRODS support for pluggable microservices. Finally, having these microservices triggered via iRODS rules releases tool developers from the burden of ensuring that tool outputs have sufficient metadata captured and are queryable. There are three microservices that add these capabilities to our iRODS storage and are explained briefly below.

Geospatial Metadata Extraction

The geospatial metadata extraction microservice is set to run whenever a new file is added to iRODS storage by attaching it to the *acPostProcForPut* event. GDAL C++ APIs are used to process the newly uploaded file to extract metadata from both raster and vector files. In particular, the title (if any), data variables, and history (if any) are extracted and stored as iRODS metadata AVU triples for the file object. In addition, the geospatial bounds of the file are extracted and converted into Lat-Long coordinates. The Dublin-core metadata schema is used to manage file metadata, restricting metadata extraction to the 15 primary Dublin Core Metadata Initiative (DCMI) fields. It is to be noted that most tools often attach useful metadata to their outputs. Past experience suggests that users are often reluctant to update file metadata if there is no pre-defined schema or if there are too many required fields. Restricting metadata to the 15 DCMI fields and attempting to automatically extract as much metadata as possible allows us to overcome these issues.

Metadata Indexing

Metadata extracted from files is also indexed into Apache Solr to enable subsequent search for files. The microservice responsible for indexing file metadata constructs an XML document from the iRODS AVU triples and uses cURL to POST this data to a remote Solr server. While this microservice is executed in conjunction with the microservice extracting metadata, a separate microservice is made available to handle user modifications from the project files web front-end. This microservice allows a set of AVU triples to be provided as input and bulk updates both the iRODS file metadata as well as the Solr index. An iRODS rule can be used to execute this microservice on demand from the HUBzero CMS. The Solr schema is designed to index all text fields into a single searchable field, allowing for file searches to conduct a simple keyword search across all fields. Any extracted geospatial bounds are indexed in a separate coverage field against which intersection queries can be conducted, enabling geospatial search from the hub. A bounding box can be drawn on a world map, and all files with data in that region can be queried and returned.

Geospatial Preview

The ability to visualize a geospatial file allows users to quickly verify that their tool (with geospatial outputs) performs as expected. More generally, users exploring geospatial datasets can gain a better picture of the data when visualized on a map. A color-coded raster map can provide an intuitive understanding of data variation across the region of interest. Overlaying multiple datasets can often provide new insights into the data. Overlaying a vector map displaying watershed regions over a raster map with crop yield data can help identify reasons for yield variance in a particular region. The GeoServer map server is used in GABBS to visualize vector and raster data. Data files are registered as layers in GeoServer and can be served using either the Web Map Service (WMS) or Web Feature Service (WFS) protocols and visualized using any mapping library that supports these protocols. The hub project files web interface, supports the preview of a subset of the raster and vector file formats using the OpenLayers Javascript library. The process of registering the raw data files into GeoServer is performed by an iRODS microservice. This microservice is responsible for converting and re-projecting geospatial files into a format handled by GeoServer. GDAL C++ APIs are used to perform the necessary transformation of the files, while cURL is used to register these files onto a remote GeoServer via its Representational State Transfer (REST) interface. While file contents can be POSTed to GeoServer when creating a layer, we instead use NFS to allow GeoServer access to the transformed files stored in the iRODS vault for efficiency reasons. Since file preview is required on-demand, the preview microservice is invoked via an iRODS rule from the HUBzero CMS.

CONCLUSION AND FUTURE WORK

By integrating iRODS storage management with the HUBzero cyberinfrastructure framework, the GABBS project accomplishes the primary goal of the DIBBS program, providing a data-centric cyberinfrastructure platform fostering collaborative research. The value-added services integrated into iRODS via the three microservices allow us to accomplish the stated goals of the GABBS project, simplifying the integration of geospatial data with disparate domains via reusable geospatial file processing, visualization and search capabilities. Going forward, we intend to expose a data service API that will allow non-hub frameworks to access data stored in iRODS for better interoperability with other DIBBS projects. We are also exploring the use of iRODS resource federation to enable distinct hubs to share data and tools with one another. This would further enable researchers from different fields to readily share expertise without having to replicate resources across their hubs.

ACKNOWLEDGMENTS

This work has been supported in part by the National Science Foundation grant #1261727. We would like to thank David Benham, Erich Heubner and Shawn Rice from the HUBzero team at Purdue University for their invaluable assistance in enabling this integration effort. We would also like to thank the iRODS team and the DataNet Federation Consortium for their continued support through this endeavor.

REFERENCES

[1] Zhao, L., Song, C.: DRINET Hub for Drought Information Synthesis, Modeling and Applications, https://hubzero.org/resources/433

[2] Villoria, N., Hertel, T.: GEOSHARE: Geospatial Open Source Hosting of Agriculture, Resource & Environmental Data for Discovery and Decision Making, **https://mygeohub.org/resources/723**

[3] Andresen, J., Jain, A. K., Niyogi, D. S., Alagarswamy, G., Biehl, L., Delamater, P., & Hart, C.: Assessing the Impact of Climatic Variability and Change on Maize Production in the Midwestern USA. AGU Fall Meeting Abstracts, Vol. 1, p. 01 (2013)

[4] Haigh, T., Takle, E., Andresen, J., Widhalm, M., Carlton, J.S. & Angel, J.: Mapping the Decision Points and Climate Information Use of Agricultural Producers Across the US Corn Belt. Climate Risk Management, 7, 20—30 (2015).

[5] https://hubzero.org/

[6] Klimeck, G., McLennan, M., Brophy, S. P., Adams, G. B., Lundstrom, M. S.: nanoHUB.org: Advancing Education and Research in Nanotechnology. Computing in Science and Engineering, 10(5), 17—23 (2008)

An R Package to Access iRODS Directly

Radovan Chytracek Nestlé Institute of
Health Sciences
EPFL Campus, Lausanne, Switzerland
Radovan.Chytracek@rd.nestle.com

Bernhard Sonderegger Nestlé Institute
of Health Sciences
EPFL Campus, Lausanne, Switzerland
Bernhard.Sonderegger@rd.nestle.com

Richard Coté Nestlé Institute of Health
Sciences
EPFL Campus, Lausanne, Switzerland
Richard.Cote@rd.nestle.com

ABSTRACT

The **R** language is an environment with a large and highly active user community in the field of data science. At NIHS we have developed the **rirods** package which allows user-friendly access to irods data objects and metadata from the **R** language. Information is passed to the **R** functions as native **R** objects (e.g. data-frames) to facilitate integration with existing **R** code and to allow data access using standard **R** constructs.

To maximize performance and maintain a simple architecture, the implementation heavily relies on the **iCommands** C++ code adapted for use with **Rcpp** and **R** language bindings.

The **rirods** package has been engineered to have semantics equivalent to the **iCommands** and can easily be used as a basis for further customization. At the NIHS we have created an ontology aware package on top of **rirods** to ensure consistent metadata annotations and to facilitate query construction.

Keywords

iRODS, R, R-language, iRODS clients, Metadata management

INTRODUCTION

iRODS[1] is a widely used open source data management solution. Some of the main features provided by **iRODS** are association of files with structured metadata, virtualization of storage with the convenience of a logical namespace and policy based automation of data management tasks. In order to take full advantage of metadata for both discoverability of data and policy enforcement, it is essential that metadata be populated for new files. This can be accomplished in different ways, depending on whether the metadata can be extracted from the file itself or determined from the context of the file creation (user, location, *etc.*). The most generic solution is to use clients which are aware of the metadata layer and are able to both populate and query it.

R[2] is a popular language and framework to perform statistical analysis on data. Scientists working in **R** require access to their data of interest, which may be available from a number of sources using differing technologies. There are many solutions in **R** for accessing structured data from databases as well as dedicated packages to read and write specific file formats, however there has been a dearth of solutions to access **iRODS**. The existing REST-based r-irodsclient[3] depends on a REST service and does not allow discovery of files via metadata queries.

We have built an integration solution whose aim is to allow users within **R** to work efficiently and comfortably, following the **R** paradigms they are familiar with. Standard **R** objects are accepted and returned. Furthermore, the design minimizes dependency on tools beyond **iRODS** itself in order to both limit complexity of deployment and to benefit from the performance optimizations present in the core **iRODS**.

iRODS UGM 2016 June 8-9, 2016, Chapel Hill, NC

R Paradigms used

Data frames

R is a language primarily targeted towards statistical analyses and is therefore well suited to handling tabular or matrix data. Various data structures exist to hold this data, but one of the most commonly used ones is the data frame. Data frames can be described as lists of named vectors of equal length. The vectors may hold elements of different types and a special vector of row names may be defined. **R** users are generally familiar with the manipulation of data frames. Wherever possible we accept data frames as input and return data frames as output.

Named arguments

R functions allow arguments to be entered using either positional syntax, named syntax or even a mixture of both. We have attempted to provide a consistent naming of arguments across different icommand-style functions.

Design Choices

A series of design choices were made to meet our specific needs. However these choices make sense in a variety of deployment scenarios.

First of all a conscious choice was made to use the **iRODS C++ client API** libraries rather than incorporating an additional component such as a REST API. Such a component introduces a performance bottleneck (all traffic is routed through the REST server rather than having clients connect directly to the appropriate resource server), adds to the complexity of deployment and constitutes an additional point of failure.

The **Jargon** api[6] was investigated as a possible solution, but **C++** was preferred mainly because it integrates much more smoothly with **R**. This does mean however, that we cannot support Microsoft Windows clients; a drawback not considered applicable in our use cases.

Many of the **iCommands** selected for implementation have been simplified in order to reduce the number of parameters they can receive. Most notably, all commands returning file and collection listings were simplified to return a single format data frame, generated through a general query. This data frame can be filtered down in **R** to select columns (or rows) of interest. Having a consistent output promotes code reuse in R.

Authentication in **R** poses an issue as **R** is often run in batch mode and providing passwords in such a use case without placing them in plain text within a password file or even within the script itself is insecure. Furthermore secure capture of the password in an interactive setting is complicated by the differences between standard **R** and **R-Studio** environments. In the end, we ask our users to perform authentication using iinit outside of **R** and use the same `.irodsA` file as the **iCommands** do. This has the added advantage of further maximizing code reuse from the **iCommands**. We do provide an insecure `iinit` function in **R** for use cases which do not allow access to iinit in the environment running **R**.

The **iCommands** implementation of the `imeta` command can be run as a subshell. This functionality has been removed since **R** allows for much more flexible scripting of complex imeta manipulation.

Due to the absence of a stable general purpose **C++ API** for much of the functionality, we have hooked directly into the **iCommands** code, simply handling argument parsing and output generation. The notable exception to this was `ils` which was re-implemented using general query as stated above.

The **C++** layer is intentionally kept simple and close to **iCommands** code. Specific **R** features such as named arguments and handling of data frames as input are implemented in the **R** layer of the package. It is trivial to customize functionality to meet site specific requirements such as metadata related functions with named arguments based on specific metadata schemas as well as metadata validation.

RESULTS
Functionality

iRODS offers many functions in its **iCommands** suite including advanced file and metadata handling commands as well as various administration commands. For **R** integration we did not consider reimplementing the whole spectrum of available functionality. We have instead implemented the minimum viable set with strong focus on the file and metadata handling in order to enable data scientists to work with **iRODS** seamlessly.

Selected features of iCommands
- iRODS connect/disconnect functions
- File level operations
- Metadata level operations
- Querying **iRODS** for files and metadata

iRODS Connection

R integration requires to establish/close **iRODS** connections and allow keeping them open during the program execution. Unlike interactive **iCommands** sessions, R sessions require batch style execution where no users are available to hanfle login/password handshake. This has impacts on the current implementation, please see the *Design Choices* section for reference. The basic set of **iRODS** connection related functions is:

$$(\text{iinit}) \quad \text{iexit} \quad \text{ienv}$$

The iinit implementation is provided for convenience. No attempt has been made to secure password entry.

File and Collection Manipulation

We have selected only some of the file related functions for **R** integration:

```
ils   iget    iput   ipwd
icd   imkdir  irm    icp
```

We have implemented the atomic `iput` with metadata and acls which has been available in **iRODS** since version 4.1.0. The `ireg` command has been left out in the current implementation.

Metadata Functions

The nature of **R** programs does not follow the original design of the `imeta` command and its sub-shell commands. We have decided to split it into individual functions instead. More details of this implementation are in the *Implementation* section.

```
imeta_add(w)   imeta_ls(w)   imeta_cp
imeta_rm(w)    imeta_mod     imeta_set
```

(w) The wildcarded arguments for these commands have been implemented

Queries

rirods implements the metadata query function. The **iCommands** command `iquery` has not been implemented. Instead we have implemented a simplified query function dedicated to metadata queries which we have called **isearch**.

```
imeta_qu   isearch
```

Implementation

The **R-package** is implemented using **Rcpp**, which automates many of the details of expanding **R** functionality with **C++** code. Basic functionality is implemented in **C++** by wrapping or modifying **iCommands** code. **Rcpp** autogenerates **R** wrappers for the **C++** functions which can then be customized as needed. This setup is shown in Figure 1.

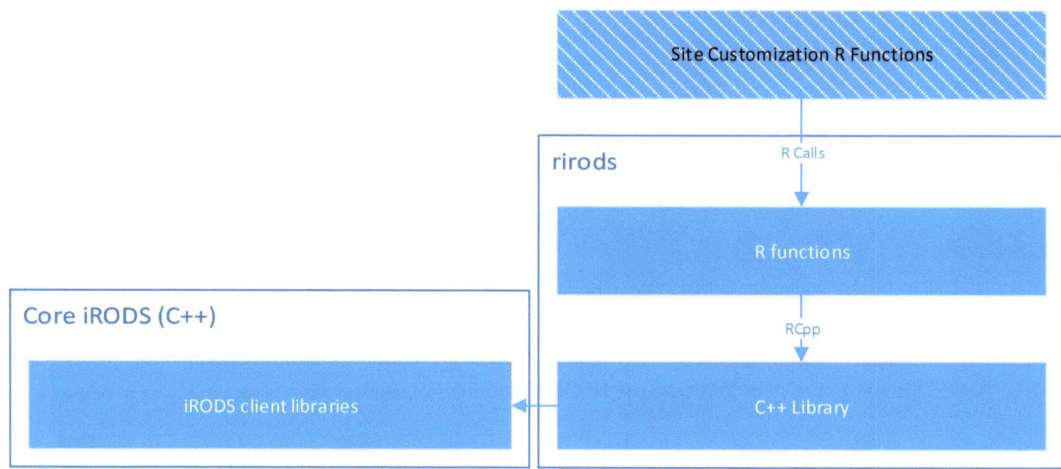

Figure 1: R and C++ components of the Rirods package. A dedicated C++ library which wraps/calls/implements iRODS interaction is accessed using the standard Rcpp mechanism. This library links to the required iRODS client api libraries. It is simple to implement customizations such as metadata validation and metadata-schema specific search and annotation functions through an additional R-package (shaded box).

The basic **rirods** package has been designed to be as generic as possible, while maintaining the usability and syntax of the original **iCommands**. We leveraged the extensible nature of **R** packages to implement an additional package which has been customized to our local needs and adds controlled vocabulary support. This **CustomRods** package is specific to our metadata schemas and standards and is therefore not released. It wraps the functions in the base **rirods** package to make them ontology-aware and enforce naming and required metadata conventions. Functions are defined using named arguments rather than generic data frames. In this context, the function calls are more intuitive for the user, who can use **R tab-completion** to quickly see what parameters are mandatory or optional. The method internals will also validate user input against an ontology file that is deployed along with the localized package. Once all inputs are validated, the generic data structures required by the base **rirods** package are constructed and passed to the low-level functions. We also provide ontology-aware lookup functions to allow users to query for valid metadata values for a given metadata attribute.

```
iput <- function(src_path, dest_path, file_type = NA, file_format = NA, software_platform = NA,
                 author = NA, relates_to = NA, ...,
                 force = FALSE, progress = FALSE, verbose = FALSE
                 )
```

Code 1: **CustomRods** implementation of `iput`. Named arguments are available for metadata fields and some arguments like checksum are not available as defaults are enforced.

The **CustomRods** call, using named parameters, in much more descriptive than the base **rirods** call, where all of the named attributes have been collapsed into a single metadata data frame.

```
iput <- function(src_path, dest_path, data_type = "", force = FALSE,
              calculate_checksum = FALSE, checksum = FALSE,
              progress = FALSE, verbose = FALSE,
              metadata = "", acl = "")
```

Code 2: **rirods** implementation of `iput`.

Run-time initialization

The default initialization pattern for the **iCommands** is to initialize global run-time variables and all plugins at each **iCommand** call. While this is perfectly normal for individual executables, the same pattern cannot be used when being run from within the **R** envionment. In this context, we need to initialize the global run-time and plugins only once and allow their re-use across the complete execution of an **R** session. In the process of developing the **rirods** package, the dynamic linker flags passed on to **Rcpp** had to be adjusted to a different shared object resolution model.

```
#include <R.h>
#include <Rinternals.h>
#include <R_ext/Rdynload.h>
#include "irods_client_api_table.hpp"
#include "irods_pack_table.hpp"

void R_init_rods(DllInfo *info) {
/* Register routines, allocate resources. */
    // initialize pluggable api table
    irods::api_entry_table&  api_tbl = irods::get_client_api_table();
    irods::pack_entry_table& pk_tbl  = irods::get_pack_table();
    init_api_table( api_tbl, pk_tbl );
}
```

Code 3: C++ iRODS shared objects initialization code executed at r-rods package loading time

```
.onLoad <- function (libpath, pkgname) {
    # We load irods shared library with global symbol resolution enforced to make sure
    # that iRODS internal plugins can also resolve symbols from iRODS API library
    library.dynam("irods", pkgname, libpath, verbose = TRUE, local = FALSE, now = FALSE)
}
```

Code 4: iRODS shared objects initialization code executed at r-rods package loading time

Examples

Simplified ils

Our `ils` implementation returns a data frame with *Data_name, Collection_name, Data_path, Data_size, Data_type, Create_time, Modify_time* for each file or collection. A generic mechanism to produce the appropriate **iRODS** general query is used for both `ils` and `isearch`. The following example shows a data frame returned from `ils`.

```
> library(rirods)
now dyn.load("/home/user/R/x86_64-redhat-linux-gnu-library/3.2/rirods/libs/rirods.so") ...
> ils()
   Data_name    Collection_name  Data_path      Data_size Data_type   Create_time         Modify_time
1               /Zone/Project-1001                    -1         C 2014-12-08 11:46:25 2015-05-26 17:56:55
2               /Zone/Project-1002                    -1         C 2014-12-04 15:52:08 2015-05-28 14:33:00
3               /Zone/Project-1003                    -1         C 2014-12-04 15:52:20 2015-05-28 14:37:26
4               /Zone/Project-1004                    -1         C 2014-11-03 13:38:58 2015-06-09 22:34:15
5               /Zone/Project-1005                    -1         C 2014-12-04 15:52:22 2015-05-28 14:33:01
6               /Zone/Project-1006                    -1         C 2014-12-04 15:52:30 2015-05-28 14:33:02
7               /Zone/Project-1007                    -1         C 2016-03-22 16:12:08 2016-03-22 16:12:08
8               /Zone/home                            -1         C 2014-10-03 15:40:41 2014-10-03 15:40:41
9               /Zone/TEST-123456                     -1         C 2015-04-16 14:55:31 2015-04-16 14:55:31
10              /Zone/trash                           -1         C 2014-10-03 15:40:41 2014-10-03 15:40:41
11 datafile     /Zone/         /iRODS/Vault/datafile 17         d 2015-03-13 13:39:47 2015-03-13 13:39:47
#Filtering for files only and a subset of columns.
> df <- ils()
> df[df$Data_type=="d",c("Collection_name","Data_name","Data_size")]
   Collection_name  Data_name Data_size
11      /NIHSData datafile        17
```

Code 5: Returning a dataframe from `ils` and manipulating it.

imeta Semantics

The examples below demonstrate the different outputs for different metadata queries. The data format is not uniform as it is the legacy of the **iCommand's** printing of different results based on the type of metadata query.

```
> # List existing metadata for test.zip
> irods::imeta_ls(type = "d", name = "/DataTest/home/rd/test.zip")
  Attribute      Value        Unit
1 FileType          ZIP
2   Purpose Test file Dummy Unit
> irods::imeta_lsw(type = "d", name = "/DataTest/home/rd/test.zip", avu = "P%;T%;z%" )
  Attribute      Value        Unit
1   Purpose Test file Dummy Unit
> irods::imeta_add(type = "d", name = "/DataTest/home/rd/test_meta.zip",  .... [TRUNCATED]
[1] 0
> irods::imeta_ls(type = "d", name = "/DataTest/home/rd/test_meta.zip")
          Attribute    Value    Unit
1 NewMetaAttribute NewValue NewUnit
> irods::imeta_mod(src_type = "d", src_name = "/DataTest/home/rd/test_meta.zip", ol .... [TRUNCATED]
[1] 0
> irods::imeta_ls(type = "d", name = "/DataTest/home/rd/test_meta.zip")
             Attribute         Value          Unit
1 ChangedMetaAttribute ChangedNewValue ChangedNewUnit
> irods::imeta_set(type = "d",
                   name = "/DataTest/home/rd/test_meta.zip", avu = "ChangedMetaAttribute;SetNewValue")
[1] 0
> irods::imeta_ls(type = "d", name = "/DataTest/home/rd/test_meta.zip")
             Attribute      Value Unit
1 ChangedMetaAttribute SetNewValue
> irods::imeta_cp(src_type = "d", dst_type = "d", src_name = "/DataTest/home/rd/tes ..." ... [TRUNCATED]
[1] 0
> irods::imeta_ls(type = "d", name = "/DataTest/home/rd/test_meta.zip")
             Attribute      Value      Unit
1             FileType        ZIP
2 ChangedMetaAttribute SetNewValue
3              Purpose   Test file Dummy Unit
> # Query iRODS objects using metadata
> irods::imeta_qu(src_type = "d", query    = "FileType = ZIP")
                 Collection          File
1            /DataTest/home/rd test_meta.zip
2            /DataTest/home/rd      test.zip
> irods::imeta_qu(src_type = "C", query    = "UserClass = Bravo")
                 Collection
1            /DataTest/home/rd
```

Code 6: Usage of imeta subcommands

Controlled Vocabulary Wrapper

The **CustomRods** package uses an `.onAttach` **R** hook to parse and load an ontology file written in the **Open Biomedical Ontology** (OBO)[7] format and distributed as external data within the package. This file contains a complete ontology that is used to describe the metadata that is allowed within our local **iRods** environment. The ontology object is stored as a session variable for rapid access throughout the **R** session. To make certain **iCommand** functions metadata-aware, we have written helper functions in **R** to confirm if values are allowed for certain attribute categories.

```
buildDataFrame <- function (attribute, value, check_valid_ontology_term=FALSE, term_category=''){

  if (missing(value)){
    renurn (data.frame("Attribute" = c(attribute), "Value" = c(NA)))
  }

  # Handle list of values
  if (length(value) > 1){
    #filter out NA
    value <- value[!is.na(value)]
    attrs = c()
    vals = c()

    for (v in value){
      if(check_valid_ontology_term){
        v = onto_check(v, check_valid_ontology_term, term_category)
      }
      attrs <- append(attrs, attribute)
      vals <- append(vals, v)
    }
    #return concatenated string of non-null values
    return (data.frame("Attribute" = attrs, "Value" = vals))

  # Handle single value
  } else {

    #ignore null/NA
    if (is.null(value)){
      return (data.frame("Attribute" = c(attribute), "Value" = c(NA)))
    }
    if (is.na(value)){
      return (data.frame("Attribute" = c(attribute), "Value" = c(NA)))
    }
    #quote valid value
    if(check_valid_ontology_term){
      value = onto_check(value, check_valid_ontology_term, term_category)
    }
    return (data.frame("Attribute" = c(attribute), "Value" = c(value)))

  }

}
```

Code 7: Function to build a dataframe of metadata attributes and values. The `check_valid_ontology_term` flag determines whether the value is verified against the ontology

```
iput <- function(src_path, dest_path, file_type = NA, file_format = NA, software_platform = NA,
                 author = NA, relates_to = NA,
                 force = FALSE, progress = FALSE, verbose = FALSE
                 ){

  #build df
  #ATTENTION, METADATA NAMES NEED TO MATCH META_DATA_ATTR_NAMES IN IRODS
  #Fields reuiring ontology check
  metadata <- buildDataFrame('File type' ,file_type,
                                           TRUE, 'File type'),
  metadata <- merge(metadata, buildDataFrame('File format',file_format,
                                           TRUE, 'File format'),
                 all.x=TRUE, all.y=TRUE)
  metadata <- merge(metadata, buildDataFrame('Software platform', software_platform,
                                           TRUE, 'Software platform'),
                 all.x=TRUE, all.y=TRUE)
  # Fields not requiring ontology check
  metadata <- merge(metadata, buildDataFrame('Author', author), all.x=TRUE, all.y=TRUE)
  metadata <- merge(metadata, buildDataFrame('Relates to', relates_to), all.x=TRUE, all.y=TRUE)

  #remove NA values
  metadata <- na.omit(metadata)
  if (nrow(metadata) == 0) {
    metadata <- NULL
  }

  return (irods::iput(src_path, dest_path, data_type = "", force, calculate_checksum=TRUE,
                      checksum=TRUE, progress, verbose, metadata, acl=""))
}
```

Code 8: **CustomRods** imlementation of `iput` with calls to the buildDataFrame function, enforcing ontology checks where needed.

In some cases, values can be free-text and not ontology-bound. In other cases, only approved values as defined in the ontology file can be used. The `onto_check` function is used to validate that a given term value is allowed for an attribute category. If the user submits a value that is not defined in the ontology, the **R** script will generate an error and stop.

CONCLUSION

The **rirods** package for the **R** language allows direct connection to **iRODS** using iCommands-like syntax and code. In order to integrate well with the **R** language, **R** paradigms such as the use of data frames and named arguments are used.

A functional subset of the **iCommands**, focussing on file and metadata management, has been implemented as **R** functions in the **rirods** package using **Rcpp** to bind **R** functions to their **C++** implementations. The **R-functions** can be further customized in order to produce site-specific functionality such as metadata validation.

Future development

While the package is functional as it is and covers the large majority of data analysis use cases, there is room for improvement. Future directions we envisage include:

- Transfer of maintenance to the **iRODS** consortium: This would allow it to be kept in close synchronization with the **iCommands** code.

- Streaming and random acces to **iRODS** files: Currently the package creates a local copy of files when `iget` is called. This can involve a large amount of I/O and network traffic. Many dedicated functions exist both in **R** and in specific libraries, which read and write from particular file formats. Most of these read from a proxy object called an **R connection object**. Implementing such an object for `iget` and `iput` would enable streaming of data directly into R. If in addition the connection object was to support random access, any optimizations in the dedicated functions to minimize data transfer volumes would be in effect.

- Secure implementation of iinit.

Availability

The **rirods** package will soon be available on gihub under the iRODS organization `https://github.com/irods`. The CustomRods package is not made available as its functionality is not transferable to other instances of **iRODS**.

ACKNOWLEDGMENTS

We would like to thank Christine Chichester for her support and the data analysts at the Nestlé Institute of Health Sciences for their valuable time and feedback during the development of the rirods package.

REFERENCES

[1] Arcot Rajasekar, Reagan Moore, Chien-yi Hou, Christopher A. Lee, Richard Marciano, Antoine de Torcy, Michael Wan, Wayne Schroeder, Sheau-Yen Chen, Lucas Gilbert, Paul Tooby, and Bing Zhu. 2010. iRODS Primer: Integrated Rule-Oriented Data System. Morgan and Claypool Publishers.

[2] R Core Team (2014). R: A language and environment for statistical computing. R Foundation for Statistical Computing, Vienna, Austria. URL http://www.R-project.org/.

[3] `https://github.com/irods/r-irodsclient`

[4] Eddelbuettel, Dirk (2013) Seamless R and C++ Integration with Rcpp. Springer, New York. ISBN 978-1-4614-6867-7

[5] `https://github.com/hadley/devtools`

[6] `https://github.com/DICE-UNC/jargon`

[7] Smith, B., Ashburner, M., Rosse, C., Bard, J., Bug, W., Ceusters, W., Goldberg, L.J., Eilbeck, K., Ireland, A., Mungall, C.J., the OBI Consortium, Leontis, N., Rocca-Serra, P., Ruttenberg, A., Sansone, S., Scheuermann, R.H., Shah, N., Whetzel, P.L., Lewis, S.: The OBO Foundry: coordinated evolution of ontologies to support biomedical data integration. Nature Biotechnology, 25(11), 1251–1255 (2007)

Davrods, an Apache WebDAV interface to iRODS

Ton Smeele
IT Services, Utrecht University
Heidelberglaan 8, Utrecht,
The Netherlands
a.p.m.smeele@uu.nl

Chris Smeele
IT Services, Utrecht University
Heidelberglaan 8, Utrecht,
The Netherlands
c.j.smeele@uu.nl

ABSTRACT

This paper covers the development and architecture of Davrods, an Apache WebDAV interface to iRODS. Our university needed a successor to the Webdavis interface that was no longer being maintained. Davrods complies with WebDAV Class 2 specifications and supports native and PAM authenticated connections to iRODS 4+ data grids. Our C language based interface connects to iRODS using the iRODS 4.1.8 client library which is backwards compatible with iRODS 3.3.1. Based on performance tests we have found that Davrods performs at least as well as its predecessor Webdavis.

Keywords

Research data management, iRODS, WebDAV, Apache mod_dav, client software, infrastructure.

INTRODUCTION

Utrecht University is an internationally prominent, research-led university that carries out fundamental and applied research across a wide range of academic fields. A high quality research data infrastructure is a precondition to its research.

Utrecht University aims to facilitate researchers to manage and share their research datasets within and across disciplines, institutes and country borders via infrastructures based on iRODS data grid technology. Easy access from researcher workstations to datasets stored in iRODS is achieved through WebDAV, while web portals allow the researcher to manage metadata and dataset integrity.

Up to and including iRODS 3.3.1, WebDAV connectivity to iRODS has been provided by a component called Webdavis. The Utrecht University project Davrods aims to develop a successor to Webdavis as a strategy to ensure long term availability of WebDAV connectivity for its iRODS based infrastructures. In this paper we describe how the architecture of Davrods supports efficient connectivity to iRODS.

ADVANTAGES OF WEBDAV

WebDAV is short for Web-based Distributed Authoring and Versioning [1]. It is a set of extensions to the HTTP protocol which allows users to collaboratively edit and manage files on remote web servers.

The WebDAV protocol provides four compelling advantages for connectivity to iRODS.

Supported standard: WebDAV is an Internet Engineering Task Force standard that has been implemented across widely used software platforms including Linux, Mac OS X and Microsoft Windows.

Connectivity: WebDAV reuses existing internet web access infrastructure over HTTPS and outgoing connections cross most organizational firewalls without any trouble. Other protocols may involve network firewall changes before they can be used by researchers.

iRODS UGM 2016, June 8-9, 2016, Chapel Hill, NC.

User experience: Access to grid-based data files and collections is conveniently transparent; it follows local workstation operating system look and feel conventions for network attached storage. Users simply drag-and-drop data back and forth using their native file manager. Local applications can access grid data without modifications.

Security: The WebDAV service can act as a gateway/stepping stone to the data grid allowing us to expose a controlled set of iRODS functions to the internet.

DEVELOPMENTS LEADING TO PROJECT DAVRODS

In 2009 Shunde Zhang [2], at the University of Adelaide, released a Java web application called Webdavis[1] that enables WebDAV clients to connect to iRODS. Since then Webdavis has been deployed at various iRODS implementations around the world including Utrecht University's YOUth project [3]. Unfortunately Zhang moved on to another position and the Webdavis code is no longer being maintained[2]. The last stable release was issued in June 2012 and does not support iRODS 4+.

Despite encouraging developments[3] we lacked a reliable solution for WebDAV connectivity to iRODS4 while we had an urgent need to upgrade our infrastructure. In October 2015 we concluded that we could not afford to wait any longer and hence we decided to develop our own WebDAV interface to iRODS using the Apache mod_dav module as a starting point.

DAVRODS DESIGN GOALS

Based on Utrecht University policies and existing service levels Davrods needs to meet the following five goals:

1. Since our community deploys a variety of platforms Davrods must comply with at least WebDAV Class 2 specifications to support all major client platforms. For example Mac OS X WebDAV clients require a Class 2 WebDAV server; otherwise write operations are disabled.
2. Performance of Davrods operations must at least match Webdavis to meet existing user expectations.
3. As we have an urgent need for the WebDAV interface, Davrods should leverage existing components to keep development time short.
4. To enable us to leverage a planned federated authentication infrastructure, Davrods should support PAM as an iRODS authentication scheme.
5. To maximize its long term use and relevance to the iRODS community at large the software should be managed and packaged as an open source product.

Webdavis provides additional features beyond WebDAV not used by our university, notably a browser oriented user interface. In contrast we shall focus Davrods development on our business needs and aim for it to be lightweight: 'to do one thing well'.

DAVRODS ARCHITECTURE AND DEVELOPMENT TECHNIQUES

We select Apache HTTPD as the underlying WebDAV server technology because Apache HTTPD has a modular, extensible architecture [4] and includes a module that serves the WebDAV Class 2 protocol [5]. This allows us to support all major WebDAV client platforms and save significant development effort. We can benefit from other Apache modules to enhance Davrods services or customize implementation. The downside of a tight integration is that we are locked in to Apache technology. Other advantages are that we already deploy Apache as a webserver in our ResearchIT infrastructure, it is open source software and Apache is one of the most popular webservers on the

---

[1] https://code.google.com/archive/p/webdavis

[2] A New Zealand initiative currently seeks to refactor this code, see https://github.com/nesi/webdavis

[3] In North Carolina USA the DICE research group develops irods_webdav. An initial release has been cut in December 2015, see: https://github.com/DICE-UNC/irods-webdav

market. To interface with the iRODS server we select the C client library [6] which also leverages the full potential of the iRODS architecture that includes amongst others advanced data transfer strategies, authentication and networking options. The iRODS C client library is the reference implementation of the iRODS API protocol. We develop Davrods in the C language to fit in with the C based Apache and iRODS interfaces.

Architectural overview

Apache HTTPD handles WebDAV requests in a modularized extensible manner.

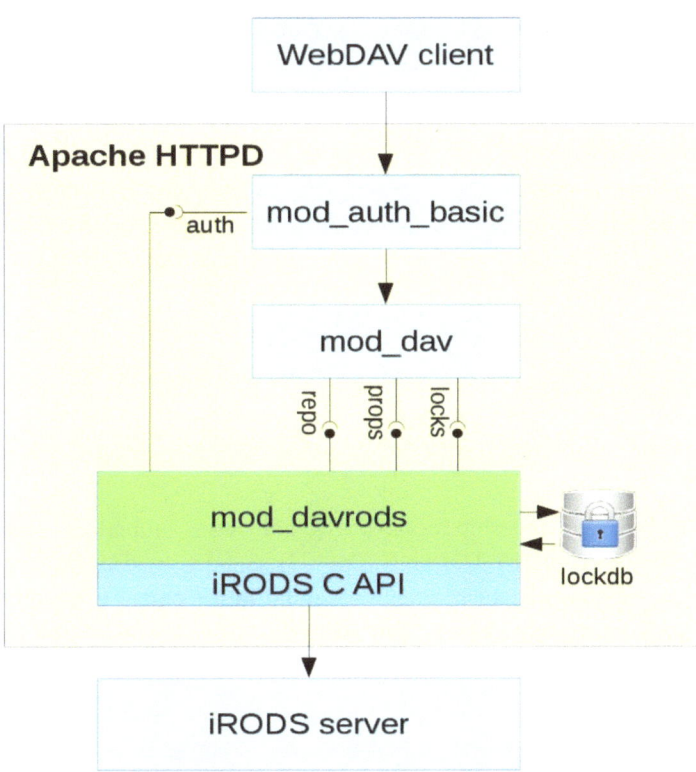

Figure 1: Davrods provides iRODS-oriented implementations for several Apache HTTPD interfaces

mod_auth_basic is an Apache module that processes HTTP headers to authenticate the client according to the basic authentication protocol. Davrods is an authentication provider to this module. We opt to use a combination of connection encryption and basic authentication rather than digest authentication as the latter is considered a less secure configuration[4].

mod_dav is the module responsible for handling DAV requests. This module parses and generates the message body of DAV requests and responses, and contains most of the logic specified by the WebDAV standard. mod_dav allows for multiple implementations of backend functionalities, called 'DAV providers'.

mod_davrods is a DAV provider that translates storage and property access requests from mod_dav to iRODS API calls and handles locking requests. In addition it is an authentication provider to mod_auth_basic.

---

[4] See note on Digest authentication in http://httpd.apache.org/docs/current/mod/mod_auth_digest.html

Session management

iRODS connections made by Davrods are strictly tied to the HTTP connection of the corresponding client. A connection is opened for each new authentication, and is maintained until either the WebDAV client or the HTTPD server (upon authentication failure or when enforcing KeepAlive restrictions) closes the HTTP connection. Technically, sessions are bound to the APR memory pool for the corresponding HTTP connection, and the iRODS disconnect is invoked by the pool's cleanup function.

We noticed that the iRODS server process queues connections upon high concurrent use. Therefore the use of prolonged open connections would have a negative impact on performance. Our approach ensures that iRODS server agent allocation time is minimized while still allowing bursts of DAV requests to reuse an open iRODS connection.

Authentication

Davrods implements a HTTPD basic authentication provider that sets up iRODS connections and performs logins via either iRODS native authentication or PAM authentication scheme. When HTTP keepalive is used, Davrods authenticates with iRODS on the first request, and for subsequent requests only when the username changes.

The DAV repository backend

The Davrods DAV repository component contains functionality for traversing iRODS collections, downloading and uploading data objects, and copying and moving objects.

The DAV property backend

The *propdb* component of Davrods is responsible for retrieving and storing properties of collections and data objects (collectively called resources in DAV speak). Currently Davrods supports read operations on DAV properties that map directly to iRODS `rcObjStat()` output (information similar to the output of a verbose `ils` command). This could be extended to allow DAV clients to access iRODS metadata.

The DAV locking backend

WebDAV Class 2 servers are required to support locking. We considered the following methods to support locking:

- *Using the iRODS locking functions*
  Making use of available locking functions in iRODS would offer interoperability with other iRODS clients. Unfortunately iRODS locking is incompatible with WebDAV locking: WebDAV allows locking collection resources, iRODS does not. Collection locks are needed for lock inheritance (with variable depth) and for gaining exclusive access to a collection's content in a single operation.

- *Using iRODS metadata*
  Storing lock bookkeeping information as metadata on the respective collections and data objects allows multiple Davrods instances to cooperate and respect each other's locks. However other iRODS clients such as icommands would not respect these locks. Properly synchronizing lock and unlock actions would involve multiple metadata operations to iRODS for a lock action on a single data object. We expect that performance would suffer greatly with this approach.

- *Using mod_dav_lock*
  Apache provides the generic DAV locking module *mod_dav_lock* which stores locking information in a local DBM database. mod_dav_lock would not support sharing locking information between Davrods instances. However, it would provide a well-tested locking solution.

Unfortunately there are several issues with mod_dav_lock that prevent us from using it directly: mod_dav_lock does not expose an interface for querying so called 'lock-null' entries. DAV repository providers depend on this functionality. Also mod_dav_lock locks resources using DAV URIs, which do not necessarily map to the same iRODS objects for different users, since the root collection of the DAV may e.g. point to the connected user's home directory.

Because it came closest to providing an acceptable solution we decided to make a fork of mod_dav_lock. Our version of mod_dav_lock is integrated into the Davrods module and resolves the issues mentioned above. We have added a lock-null query feature and we have adapted the provider to use canonical iRODS paths instead of DAV URIs.

Configuration

The behavior of Davrods is necessarily controlled by two configuration files.

Firstly, Davrods is configured just like other Apache modules using a file located in the Apache configuration directory (`/etc/httpd/conf.d`). It follows Apache configuration conventions and it allows setting iRODS connection parameters such as server location details, default resource and authentication scheme. The authentication scheme can be set to either native or PAM. Furthermore, for ease of use, Davrods can be configured to expose only a subtree of the iRODS virtual filesystem to the WebDAV client. Other directives are available to influence file upload completion behavior and to tune the size of send/receive buffers used for file transfer to/from iRODS.

Secondly, an `irods_environment.json` file is required[5] by the iRODS client library that Davrods depends upon. This file follows iRODS client configuration conventions. Advanced iRODS protocol options can be configured here such as parallel data transfer settings, SSL options and parameters that influence backwards compatibility with iRODS3 grids. Some configuration parameters such as zone name can be defined in both files. In such cases the definition in the Davrods configuration file takes precedence.

CHALLENGES

Unfortunately the iRODS C client library is not well documented. To discover appropriate library functions we consult source code including examples from other client implementations such as icommands and Kanki[6] [7]. Memory management responsibilities related to library function call parameters are not always clear. In those cases we assume the intended behavior and document our concerns inline.

Our initial development environment is a Linux CentOS 7 platform with Apache 2.4, Apache Portable Runtime 1.4.8 and iRODS 4.1.4. The Davrods binary, dynamically linked against iRODS 4.1.4 client libraries, did not work in an environment with iRODS 4.1.8 client libraries due to an added dependency on the irods_client_plugins library. To resolve this issue we had to relink against iRODS 4.1.8 client libraries.

While we expected the original *mod_dav_lock* code to be stable we discovered (and were able to fix after many troubleshooting hours) a serious bug that caused lock database corruptions.

---

[5] We must admit that the iRODS client library version 4.1.8 allowed us to refer to a non-existing file in which case it would use default values.

[6] https://github.com/ilarik/kanki-irodsclient

PRODUCT ASSESSMENT

Support of major client platforms

Davrods passes the Litmus WebDAV server protocol compliance test [8] which asserts that our WebDAV server meets the WebDAV Class 2 specifications according to IETF RFC2518. Compatibility tests have been executed on target platforms using popular WebDAV clients: on Linux we have executed tests using davfs2, cadaver, and curl; on Mac OS X we have tested using Finder; on Windows 7 we have tested with NetDrive and Cyberduck.

Performance benchmark

We have compared performance of our Davrods based configuration with our old Webdavis based configuration to test if and how the configuration change would impact user experience.

One web application server is configured with Apache HTTPD, Tomcat and Webdavis on a RHEL 6 system. This configuration mimics our old web application server environment. Another web application server in the same local network is configured with Apache HTTPD and Davrods on a CentOS 7.2 system to mimic our new web server environment. Both web servers access the same iRODS server via the local network using the native authentication scheme. The iRODS server is iRODS3.3 on a RHEL 6 system with a local ICAT database. All hosts are virtual servers just like our production environments. We compensate for interference by other activities by running the tests repeatedly in various time slots.

Each of our tests comprises a single WebDAV session. During a session we either transfer a single 1 GB file or we transfer 500 files of 2 MB each using cadaver as a WebDAV client. During a session we either upload or download files, we do not mix them. We also test uploading and downloading a 1GB file using curl as an alternative WebDAV client.

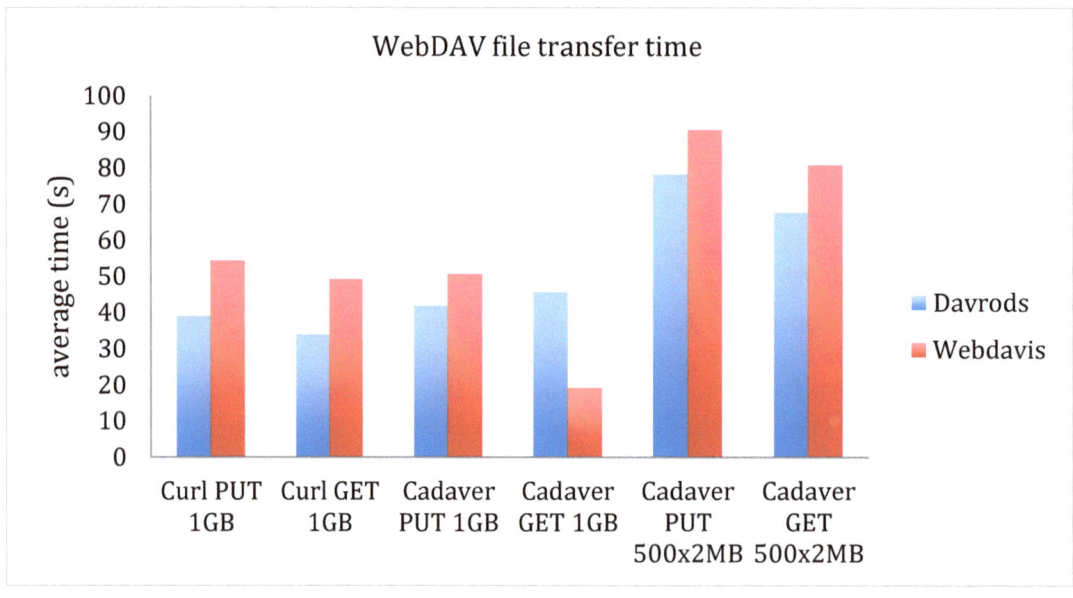

Figure 2: Comparison of file transfer performance between Davrods and Webdavis

Davrods outperforms Webdavis by 15-30% for nearly all of the above file transfer operations. However Webdavis performance beats Davrods by more than a factor two upon download of a single 1GB file when combined with cadaver as the WebDAV client. Analysis of the HTTP protocol shows that cadaver incudes headers in the web client HTTP request to indicate acceptable response formats that curl does not add. Further investigation might help us to improve Davrods performance in similar situations.

CONCLUSION AND FUTURE WORK

The fast and successful realization of Davrods provides us with a successor to Webdavis and it has enabled us to migrate from iRODS3 to iRODS4 in Q1 2016. Davrods is now used in all our iRODS grids. Researchers much like the WebDAV access to iRODS for its ease of use and cross-institute/country potential. Future developments of Davrods will focus on performance (the benchmark results indicate that there is potential for faster file transfers) and support for future iRODS server releases.

REFERENCES

[1] Internet Engineering Task Force, https://tools.ietf.org/html/rfc4918, Visited last on 13.05.2016

[2] Zhang, S., Coddington, P.D., Wendelborn, A.L.: Davis: A generic interface for iRODS and SRB. GRID. October, pp. 74--80 (2009)

[3] Consortium on Individual Development, http://www.individualdevelopment.nl, Visited last on 13.05.2016

[4] Apache Modeling Project, http://www.fmc-modeling.org/category/projects/apache/amp/3_3Extending_Apache.html, Visited last on 13.05.2016

[5] Apache Software Foundation, http://httpd.apache.org/docs/current/mod/mod_dav.html, Visited last on 13.05.2016

[6] iRODS Documentation, https://docs.irods.org/master, Visited last on 13.05.2016

[7] Korhonen, I., Nurminen, M.: Development of a native cross-platform iRODS GUI client. In: 7th iRODS User group Meeting, pp. 21—28. iRODS Consortium, Chapel Hill (2015)

[8] WebDAV Resources, http://www.webdav.org/neon/litmus, Visited last on 13.05.2016

NFS-RODS: A Tool for Accessing iRODS Repositories via the NFS Protocol

**D. Oliveira, A. Lobo Jr.,
F. Silva, G. Callou, I.
Sousa, V. Alves, P.
Maciel**
Center for Informatics
UFPE, Recife, Brazil
{dmo4,aflj,faps,grac,isf2,
valn,prmm}@cin.ufpe.br

Stephen Worth
EMC Corporation
Massachusetts, U.S.A.
stephen.worth@emc.com

Jason Coposky
iRODS Consortium
Chapel Hill, U.S.A.
jasonc@renci.org

ABSTRACT

Data center and data evolution have been dramatic in the last few years with the advent of cloud computing and the massive increase of data due to the Internet of Everything. The Integrated Rule-Oriented Data System (iRODS) helps in this changing world with virtualizing data storage resources regardless the location where the data is stored. This paper presents a tool implemented for accessing iRODS repositories through the NFS protocol. This tool integrates NFS to the iRODS server through common operating system commands on a remote iRODS repository via the NFS protocol.

Keywords

iRODS, storage, Network File System

1. INTRODUCTION

The data center has evolved dramatically in recent years due to the advent of the cloud computing paradigm, social network services, and e-commerce. This evolution has massively increased the amount of data to be managed in data centers. In this context, the Integrated Rule-Oriented Data System (iRODS) has been adopted for supporting data management. The iRODS environment can virtualize data storage resources regardless of the location where the data is stored as well as the kind of device the information is stored on.

IRODS is an open source platform for managing, sharing and integrating data. It has been widely adopted by organizations around the world. iRODS is released and maintained through the iRODS Consortium [2] which involves universities, research agencies, government, and commercial organizations. It aims to drive the continued development of iRODS platform, as well as support the fundraising, development, and expansion of the iRODS user community. iRODS is supported by CentOS, Debian, and OpenSuse operating systems.

Network File System (NFS) is a distributed file system originally developed to share data (e.g., files or directories) between computers connected through a network, which forms a virtual directory locally represented for users. In order to increase the NFS utilization, this work proposes a tool for accessing data repositories (e.g., iRODS). In order to accomplish this, the developed tool, named NFS-RODS, transparently integrates the NFS protocol with the iRODS data repositories, i.e., users are able to execute the common file system operating commands, but all the operations are remotely performed accessing the iRODS repositories through NFS. This new functionality provided by the implemented tool is presented in this paper.

This paper is organized as follows. Section 2 introduces the basic concepts needed for a better understanding of this work. Section 3 presents the developed tool named NFS-RODS as well as shows some examples related to its use.

iRODS UGM 2016 June 10-11, 2016, Chapel Hill, NC

Section 4 concludes the paper and makes suggestions on future directions.

2. DISTRIBUTED DATA SHARING

This section presents basic concepts of this work. First, the Network File System (NFS) protocol is detailed, which is followed by the integrated Rule-based Data management System(iRODS). This work combines these two concepts, in which both similarities and differences must be emphasized. The NFS protocol is an open standard defined in the following RFCs [8] [4], that provides a distributed file system. A distributed file system consists of a shared storage accessed via a network protocol, that provides an interface with the same semantics of a local file system. On the other hand, the iRODS consists in a Storage Resource Broker: a middleware for a data grid that provides a single logical namespace for a set of heterogeneous storage systems, that can span across multiple administrative domains.

In short, NFS is a standardized network protocol that allows us to treat a remote folder as a local one, and iRODS is a data-grid middleware that provides management, sharing, publication, and long-term preservation of distributed data [7]. They are different concepts that provide the same basic functionality: accessing remote files in a client-server architecture. In the next sections, we will examine in more details each of them.

2.1 Network File System

The primary goal of a NFS client is to turn the remote access transparent for the computer users. To accomplish this, the NFS adopts the client-server interface, in which the user can request a file present on the server as it was locally stored. Its interface is public and widely used for the sharing of readings and academic organizations, due to its benefits such as transparency; command unification; reduction of local space; independent of operating systems and hardware.

For a client-server system, once logged in, the client can automatically import the directories and files previously created in the personal area. However, to implement this file import system associated with one particular user, a system with LDAP must have been configured in addition to NFS.

2.2 iRODS

iRODS has become a powerful, widely deployed system for managing a significant amount of data that requires extendable metadata. Typical file systems provide only limited functionality for organizing data and a few (or none) for adding to the metadata associated with the files retained. Additionally, file systems are unable to relate or structure what limited metadata is available and provide only a platform from which to serve unstructured file data. Within several fields, scientific research evolving instrumentation capabilities have vastly expanded the amount and density of unstructured file data, in which standard file systems can be a limiting factor in the overall use of data.

iRODS can be classified as a data grid middleware for data discovery, workflow automation, secure collaboration and data virtualization. The middleware provides an uniform interface to heterogeneous storage systems (POSIX and non-POSIX). iRODS lets system administrators roll out an extensible data grid without changing their infrastructure and accessing through familiar APIs. The reader should refer to [6] and [5] for more details about iRODS environment.

3. NFS-RODS

The NFS-RODS server is an implementation of the NFS protocol that exports folders located at an iRODS server to the clients. It allows the clients to use a NFS client to access iRODS folders, typically by using the mount command and treating the remote folder as a local one.

This section explains the architecture of the NFS-RODS server, and present some details about the implementation. Then, the next section presents some examples of its utilization.

System Architecture

Figure 1 depicts the NFS-RODS tool architecture. As previously mentioned, the main goal of this tool is to provide access to iRODS repository through the NFS protocol. To accomplish this, a client-server system architecture is adopted. In the client, a local folder must be mounted into the NFS-RODS Server. Besides that, the NFS-RODS server folder must be exported to allow iRODS iCAT-Server to be able to write and read through the NFS protocol. More details about how it works is provided in the utilization example section.

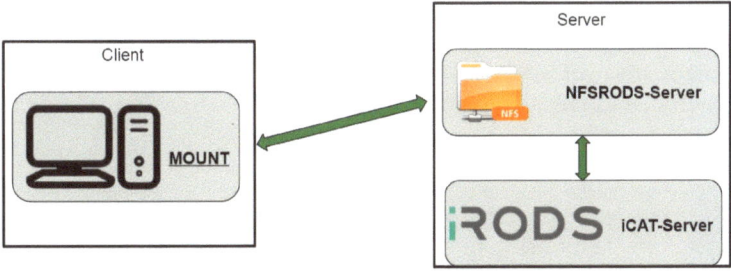

Figure 1. System Architecture.

Package Diagram

Figure 2 presents the structure of the NFS-RODS implementation as a UML package diagram. It shows the folders used to organize the source files, and the main C files of the project (the header files are omitted).

The *daemon.c* file holds the main function. Once running the NFS-RODS server, that function is called and the RPC server is started as a background process. The callbacks functions are implemented in the *nfs.c* file, which is registered on the RPC server and implements the specification described at the RFC 1813 [4]. Those functions correspond to the basic file/folder operations like: read directory (READDIR), remove file (RM), remove directory (RMDIR), read file (READ), etc. All those callbacks must operate on the iRODS file system. Therefore, the callbacks perform calls to the functions present on the iRODS C API [1] in order to allow the client to access the iRODS server via the NFS protocol.

For the sake of modularity, and for avoiding the code repetition for opening iRODS connections, we provide wrappers files for performing those basic operations (readdir, rm, rmdir, etc.), which is the *irodsapi.c* file. Other general utility functions are listed on the *utils.c* file (e.g. for debugging, and handling paths strings).

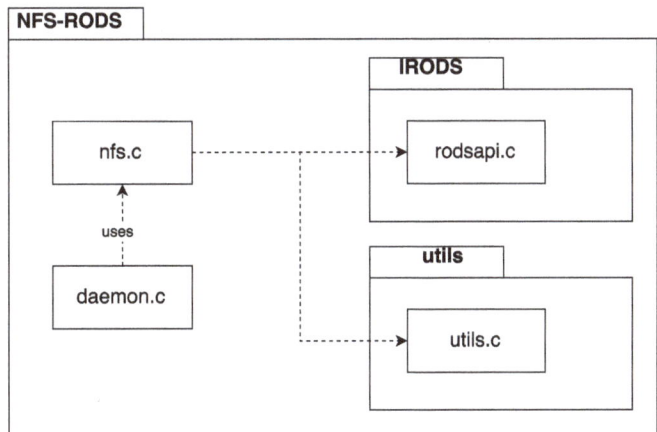

Figure 2. Package Diagram.

Utilization examples

To to be able to use our NFS-RODS tool, the user can download and install the source code, that is available at [3].

Executing using the source files

To use the NFS-RODS source code, users must download the source files and use a tool as the QTCreator. After compiling the source code, the user must execute the project. This project works as a daemon which carries out the requests that the NFS receives. In addition, users must have mounted a folder from the client into the NFS server. For instance, the command *mount serverName:/serverFolder /clientFolder*, mounts the specific folder present on the client (e.g., *clientFolder*) on a specific folder on the server (e.g., *serverFolder*). Figure 3 shows an example used for mounting a server folder running on the same machine than the client.

Figure 3. Mounting example.

Once the mount command has been executed, users can test this operation success by executing the *df* command presented on Figure 4. This shows that the local client folder has mounted on the server folder, which is also the local machine for this example. Additionally, the export file must be set to allow the iRODS iCAT server to read and write using the NFS protocol. This is configured in the /etc/exports file, in which the following line must be present: */serverFolder serverName(rw)*.

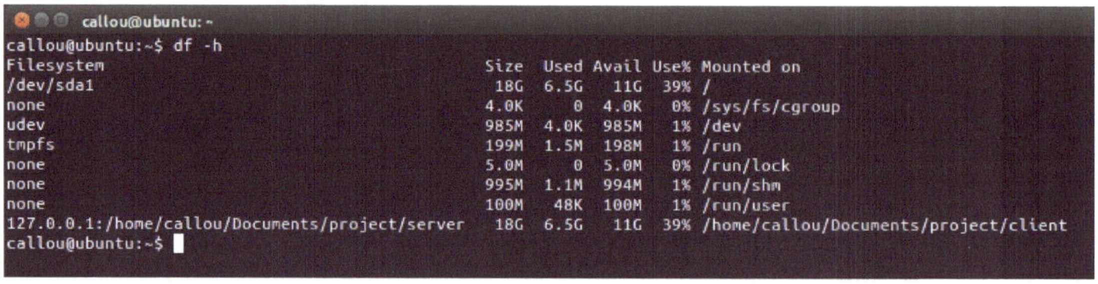

Figure 4. Checking the mount operation.

Examples

This section shows some examples to illustrate that the NFS-RODS tool utilization corresponds to the usual common operations for manipulating files on operating systems. However, it is important to stress that instead of dealing with the local data, the data has been remotely stored, and is accessed via NFS protocol. Figure 5 depicts the examples through three commands: *mkdir*, for creating a folder; *rm*, for removing a folder; and, *mv* for renaming also a folder. The reader should notice that all these commands are exactly the same ones used for manipulating local data.

To show the creating process example (Figure 5 (a)), we first executed the ils command to list the data content of our iRODS repository, which corresponds to *exemp1*, *exemp3*, *exemp4* and *exemp8* folders. Therefore, after executing the *mkdir exemp0*, we also needed to list the new content of our repository on the iRODS server to show that the *exemp0* folder was created.

Similarly to the previous example, to show the remove directory operation (Figure 5 (b)), we first executed the ils command to list the data content of our iRODS repository, which corresponds to *exemp0*, *exemp1*, *exemp3*, *exemp4*

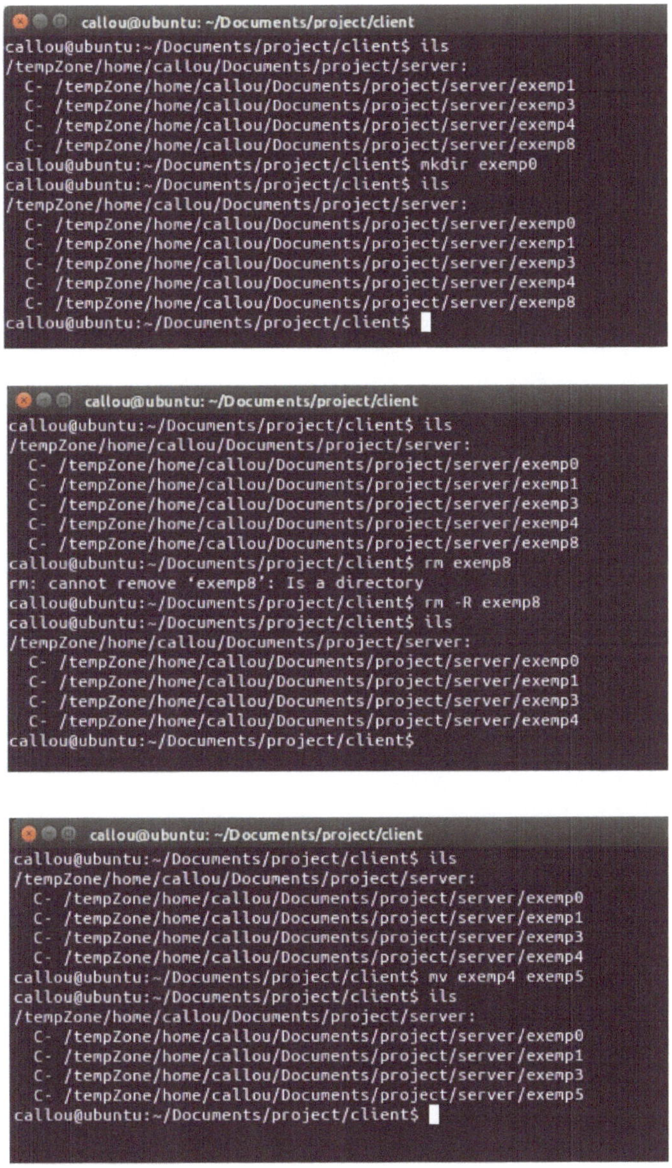

Figure 5. NFS-RODS utilization examples: (a)creating a folder - mkdir, (b) removing a folder - rm, (c) renaming a folder - mv.

and *exemp8* folders. Therefore, after executing the *rm -R exemp8* command, we also needed to list the new content of our repository on the iRODS server to show that the exemp8 folder was deleted.

Finally, the last example shows the rename operation (Figure 5 (c)). Again, we first executed the ils command to list the data content of our iRODS repository, which corresponds to *exemp0*, *exemp1*, *exemp3* and *exemp4* folders. Therefore, after executing the *mv exemp4 exemp5* command, we also needed to list the new content of our repository on the iRODS server to show that the *exemp* has been renamed for *exemp5* folder.

4. FINAL COMMENTS

The evolution of data center due to the advent of cloud computing as well as the Internet of Everything has been increasing the amount of data to be managed by that systems. The iRODS helps in this changing world with virtualizing data storage resources regardless the location where the data is stored. This work has presented and demonstrated the NFS-RODS tool for accessing iRODS repositories through the NFS protocol. As a future direction, we are extending the proposed approach to consider other data repositories (e.g., Amazon S3).

REFERENCES

[1] irods c apis documentation. `https://wiki.irods.org/doxygen_api/html/index.html`, 2012. Last access in: 2016-02-25.

[2] Irods consortium web page. `http://www.irods.org/`, 2015. Last access in: 2016-02-25.

[3] Nfs-rods: A tool for accessing irods repositories via the nfs protocol, 2016.

[4] B. Callaghan, B. Pawlowski, and P. Staubach. Nfs version 3 protocol specification. Technical report, RFC 1813, Network Working Group, 1995.

[5] iRODS. Using an integrated rule-oriented data system (irods) with isilon scale out nas. http://www.emc.com/collateral/white-papers/h13232-wp-irodsandlifesciences-isilonscaleoutnas.pdf. Acessed: 04/25/2015.

[6] iRODS. The integrated rule-oriented data system (irods). http://irods.org/, 2014. Acessed: 04/25/2015.

[7] A. Rajasekar, R. Moore, C.-y. Hou, C. A. Lee, R. Marciano, A. de Torcy, M. Wan, W. Schroeder, S.-Y. Chen, L. Gilbert, et al. irods primer: integrated rule-oriented data system. *Synthesis Lectures on Information Concepts, Retrieval, and Services*, 2(1):1–143, 2010.

[8] S. Shepler, B. Callaghan, D. Robinson, R. Thurlow, C. Beame, M. Eisler, and D. Noveck. Rfc 3530: Network file system (nfs) version 4 protocol. *IETF, April*, 2003.

Academic Workflow for Research Repositories Using iRODS and Object Storage

Randall Splinter, Ph.D.
DataDirect Networks
238 Serenoa Drive
Canton, GA 30114
RSplinter@ddn.com

ABSTRACT

Traditionally, the sharing and retention of research data has been a contentious issue. Sharing data over WANs has been limited by the available storage technologies. NAS solutions while excellent for sharing data over a LAN have never had the same success over WANs. The successful implementation of object storage solutions has opened a door into the ability to share data over WAN links.

Coupling that ability to share objects over a WAN with middleware like iRODS provides the research community with the ability to provide to provide more stringent controls over the data including

- Better control of ACLs including
    - Implementing data retention policies to meet regulatory requirements
    - Loss of IP due to faculty loss
- Virtualization of multiple storage silos under a single namespace
- Extensive metadata tags and searching of those tags
- Extensible rules engine to implement functionality such as
    - HSM style functionality between storage devices
    - Data migration based upon set criterion

Some of the advantages of this approach include

- Ease of administration – Once rules are tested and in place the system can be managed with a minimum of administrative overhead
- Automating workflows to guarantee consistency and reproducibility in the science that is produced
- Ease of auditing for both usage and back charging and for maintaining adequate data security compliance
- Using storage platforms like DDN WOS remote replication becomes simple and provides a straightforward way to manage DR systems

Keywords

Data Archives, Data Repositories, iRODS, Object Storage, WOS,

iRODS UGM 2016, June 10-11, 2016, Chapel Hill, NC.

INTRODUCTION

Over the past few years as government funding agencies have begun to adopt requirements for data management plans for research, universities and federal laboratories have struggled with the development of programs to provide long-term storage of research data. The problem is not new per se, but the recent development of object storage has finally provided an interesting solution by which data can be reliably stored long-term while providing solutions to

1. Data security from accidental deletion (retention policies), loss due to theft or hardware failures.
2. Satisfying regulatory restrictions such as HIPPA and FIPS-140-2
3. Changing hardware standards
4. Hardware availability

Historically, satisfying all of these simultaneously has been a challenge from traditional NAS and Fibre Channel storage technologies.

In this brief we would like to summarize the existing state of affairs and show how a combination of object storage and iRODS can provide a solution to long-term archiving and active research data repositories that is significantly more cost effective and flexible than more traditional methodologies.

INTRODUCTION TO THE PROBLEM

Reducing the problem to its bare essence we are left with the problem of collaboration and research repositories. Both have their challenges using traditional storage technologies. In the next two sections we wish to outline the problem in some detail before moving on to a modern solution to both problems.

The Problem of Collaboration

The first significant attempt at providing an affordable solution to the problem of collaborative storage was provided by Sun Microsystems in 1984 in the form of the Network File System (NFS)[1]. This was followed by the development of Server Message Block (SMB) by IBM around 1990[2][3]. In 1996 Microsoft renamed SMB to Common Internet File System (CIFS)[4]. Both of these technologies still enjoy a good measure of success in the marketplace, but they also have their share of problems. The biggest single obstacle is a lack of interoperability between NFS and CIFS, but they also suffer from being primarily local technologies in the sense that they do not scale well over WAN-type of distances. This significantly limits their usefulness for collaborations that are over substantial distances such as we are seeing today.

Therefore, the problem of collaboration reduces to one of finding the appropriate technology to enable the sharing data over large distances. Two interesting solutions to this problem from a software perspective have emerged over the last few years. The first is GlobusOnline[5] and the second is iRODS[6]. The GlobusAlliance has focused on the ease of transferring data over arbitrary distances. The development of GridFTP has made that the transfer of data more efficient by enabling parallel IO streams and enhancing the security of the data transfer. The iRODS team has focused on providing solutions to

1. A virtual namespace spanning all available storage resources.
2. Providing mechanism by which end users can easily modify ACLs and permissions to share data.
3. Providing mechanism by which end user can easily add metadata tags to objects and search that metadata
4. Providing a rules engine which can be used to automate workflows and simplify the movement of data based on rules

These four rules can be graphically summarized as a "Swiss Army" knife, with each blade representing one of the above items

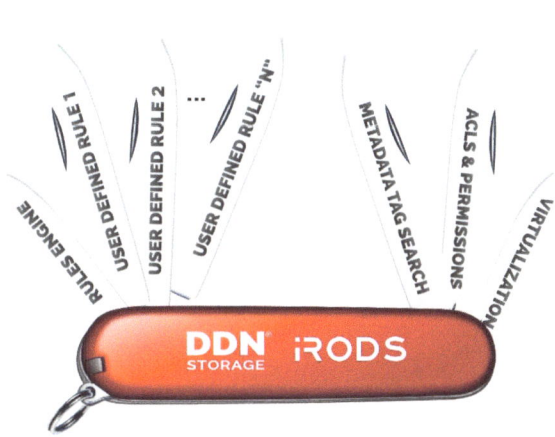

For the rest of this paper we will focus on iRODS. The Globus team is developing very interesting technology to enable data transfers, sharing, and the publishing of data. But, is beyond the scope of this work.

The Problem of Research Repositories

As commented above as government agencies have started to demand data management plans as part of grant funding the development of adequate solutions to the long-term storage of data has become more pressing. A brief list of the problems that are encountered can be summarized as

1. Insure long-term availability of the data
    a. Hardware availability/failure
    b. Changing hardware standards
2. Data security
    a. Retention policies are needed
    b. Loss due to theft – loss of IP
3. Regulatory restrictions
    a. HIPPA, FIPS-140-2, etc.
4. Enable ease of data sharing

Each of these presents their own set of problems, and frequently they can contradict one another. For instance, the need to be able to share data or research results may well provide issues for data security. External users may need access to some objects, but not to all objects. Or as another example, should a researcher leave one institution they may have a need to take some data with them, but their former employer may view that data as institutional IP.

OBJECT STORAGE AS A BRIDGE

The first proposal for object storage was given by Mesnier, et al[7]. Since then a number of competing object storage technologies have been released into the market, for example

1. DDN WOS
2. Scality RING
3. IBM Cleversafe
4. OpenStack SWIFT
5. CEPH
6. Amazon S3
7. Amplidata
8. EMC Atmos
9. Caringo
10. Plus others……

Some of these are Open Source products while most of them are commercial platforms. Nonetheless, all of them share some common features that make them ideal for the user in collaborative/repository storage.

1. In order to scale filesystems to very large scale it has been realized that POSIX locking can be a significant hurdle. Therefore, all object store technologies have abstracted away the underlying POSIX filesystem.
2. In order to avoid the POSIX issues above all limit the number of operations that an end-user can exercise, usually some combination of create, read, update and delete (CRUD).
3. Rich custom metadata for an object. This is allows end users to add metadata tags to objects and search metadata at a bare minimum.
4. The use of standards such as ReST.
5. The use of data replication. The effectiveness of replication over WAN distances depends largely on the object storage platform.
6. Erasure encoding to provide data security (in place of RAID)

HOW IRODS ENABLES OBJECT STORAGE

The role of research repositories and archives will become increasingly important over the next few years as funding agencies push for increased openness in supported research. This provides the ideal environment for bringing together object storage and iRODS.

Object storage is ideal in a Write Once, Read Many (WORM) environment. By removing the limitations imposed by POSIX locking object storage enables the growth of storage environments to massive sizes. Furthermore, the native replication techniques that object storage platforms implement provides data protection for extended periods of time. Further, some object storage platforms, such as DDN WOS, also implement erasure encoding at the local level to provide additional data protection.

On the other hand, object storage platforms in their raw form tend to be difficult to use. Most support the ReST interface natively, but the use of ReST can be problematic for less sophisticated users. Gateways, on the other hand regardless of the protocol they support (NFS, CIFS, S3, Swift, to name a few) act to provide a simple and often POSIX like interface to the raw object storage. That is somewhat counter intuitive since a great deal of work has gone into removing POSIX from the object storage in the first place. The solution, of course, is software.

iRODS stands as the best alternative for a software stack to layer on top of an object storage platform and to provide the best user experience in terms of usability. Recalling from above

1. A virtual namespace spanning all available storage resources.
2. Providing mechanism by which end users can easily modify ACLs and permissions to share data.
3. Providing mechanism by which end user can easily add metadata tags to objects and search that metadata
4. Providing a rules engine which can be used to automate workflows and simplify the movement of data based on rules

The first item allows for the inclusion of multiple storage platforms into the overall solution. This can simplify data movement between storage resources. For instance, in a HPC environment the output of compute jobs can be easily moved either by hand or through the use of the Rules Engine from high speed filesystems to the long-term object storage. Metadata tags can be added to the dataset prior to moving the data to the object storage platform to accommodate later searches. Finally, the end user can modify ACLs or create tickets to enable the straightforward sharing of data with colleague on or off campus.

AN EXAMPLE IMPLEMENTATION OF A DDN WOS WITH IRODS REPOSITORY

The following is an example of a research data repository that has been recently delivered to the Texas Tech University. The system uses DDN WOS storage as the object storage backend and iRODS as the software layer providing the needed functionality to make the repository functional.

Some critical features of the repository are

1. Two WOS zones for replication of all data between two geographically separated data centers, and local erasure encoding to provide additional data safety.

WOS Policy Name	WOS Zone	Description
wosresUDC	UDC Only	Local Object Assure with one copy at UDC
wosresRDC	RDC Only	Local Object Assure with one copy at RDC
wosresREPL	UDC and RDC	Local Object Assure at both sites with one copy at both UDC and RDC

2. A highly available environment using Corosync and Pacemaker to provide an active/passive HA cluster for the PostgreSQL/iCAT database, HAProxy services and the iRODS Cloud Browser.
3. A scale-out environment using WOS back-end storage to enable simple capacity upgrades. The simplicity of adding more WOS capacity is a critical feature of the overall design. The customer did not want to perform extensive reconfiguration after adding more capacity.
4. Multiple ingest methods including, iCommands, iRODS Cloud Browser and the GridFTP/iRODS connector. Future work will include adding a 10GbE link to the campus HPC Lustre filesystem to provide resource to move data directly into the back-end WOS storage from the HPC Lustre filesystem.
5. Authentication must use the PAM and Kerberos in order to tie into the existing campus-wide eRaider authentication system that is used campus wide.

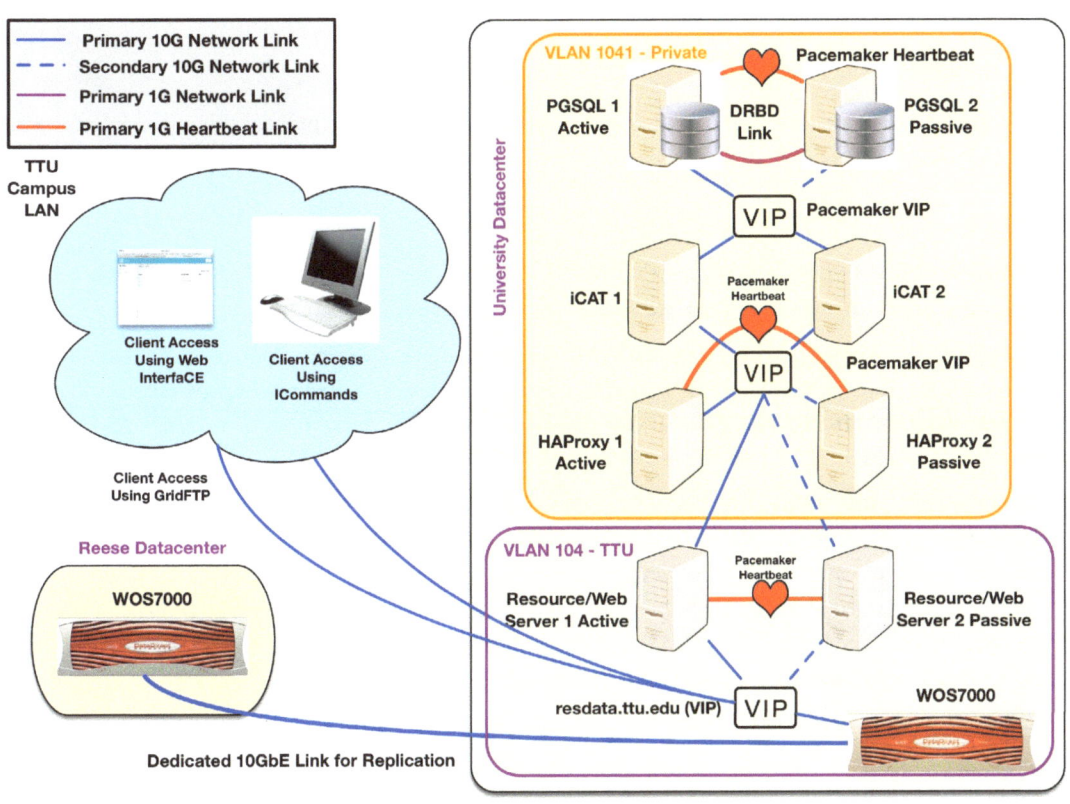

The complete solution will provide a Research Repository for TTU that will scale into the foreseeable future and provide campus researchers a long-term storage platform for the storage of long-term research data. The system has been opened for beta customers and upon completion of beta testing the system will be opened for campus wide use later this year.

It also demonstrates the flexibility and robustness that can be built into an iRODS deployment. The use of HA clustering enables resistance to downtime from hardware or software failures, while employing a scale-up and scale-out design provides a WOS cluster that scales out simply and the use of multiple iCAT servers to handle any peaks in work load that may occur during normal usage.

CONCLUSION

In conclusion we have presented arguments for the use of iRODS as a middleware stack with object storage as the physical hardware storage for use in long-term archives or repositories. By combining the two we believe that we have demonstrated through our work at TTU that it is possible to build a highly available and scalable iRODS/WOS environment.

The choice of iRODS is simple, the combination of a virtualized namespace, end-user ACLs and permissions, metadata tagging and searching and a highly extensible rules engine make iRODS the ideal middleware stack for use in an archive/repository environment. Using the iRODS rules engine, data can be migrated between multiple iRODS storage resources transparently and provide data retention policies for data that cannot be lost to accidental deletion or theft.

The ability of object storage to scale to very large numbers of Petabytes, with local erasure encoding and replication provides an ideal method for the long-term storage of data, while keeping the data secure and safe from hardware failure. The DDN WOS hardware platform was used in the example here. A WOS cluster scale out easily in terms of capacity so growth can be handled without extensive changes to the existing environment. Finally, WOS has flexibly storage policies so virtually any situation can be accommodated, from local erasure encoding to replication of objects between multiple WOS zones in the cluster.

ACKNOWLEDGMENTS

The author would like to thank the iRODS consortium for the opportunity to present this material at the 2016 iRODS User's Group Meeting.

REFERENCES

[1] *Russel Sandberg, David Goldberg, Steve Kleiman, Dan Walsh, Bob Lyon (1985).* **"Design and Implementation of the Sun Network Filesystem".** *USENIX.*

[2] **"Common Internet File System".** *Microsoft TechNet Library. Retrieved 2013-08-20. The Common Internet File System (CIFS) is the standard way that computer users share files across corporate intranets and the Internet. An enhanced version of the Microsoft open, cross-platform Server Message Block (SMB) protocol, CIFS is a native file-sharing protocol in Windows 2000.*

[3] **"Microsoft SMB Protocol and CIFS Protocol Overview".** *Microsoft MSDN Library. 2013-07-25. Retrieved 2013-08-20. The Server Message Block (SMB) Protocol is a network file sharing protocol, and as implemented in Microsoft Windows is known as Microsoft SMB Protocol. The set of message packets that defines a particular version of the protocol is called a dialect. The Common Internet File System (CIFS) Protocol is a dialect of SMB. Both SMB and CIFS are also available on VMS, several versions of Unix and other operating systems.*

[4] **Tridgell, Andrew.** *"Myths About Samba". Retrieved 2016-01-03.*

[5] **Globus Alliance established as international consortium to advance Globus grid software"** *(Press release). Globus Alliance. September 2, 2003. Retrieved 2007-08-08. the Globus Project*

today transformed itself into the "Globus Alliance." ... The Globus Project was established in 1995 by the U.S. Argonne National Laboratory, the University of Southern California's Information Sciences Institute (ISI) and the University of Chicago (UofC).

[6] *Conway, Mike; Moore, Reagan; Rajasekar, Arcot; Nief, Jean-Yves (2011). "Demonstration of Policy-Guided Data Preservation Using iRODS". Proceedings of the 2011 IEEE International Symposium on Policies for Distributed Systems and Networks: 173–174.* **doi:10.1109/POLICY.2011.17. ISBN 978-0-7695-4330-7***.*

[7] *Mesnier, Mike; Gregory R. Ganger; Erik Riedel (August 2003).* **"Object-Based Storage"** *(PDF). IEEE Communications Magazine: 84–90.* **doi:10.1109/mcom.2003.1222722***. Retrieved 27 October 2013.*

Application of iRODS metadata management for cancer genome analysis workflow

Lech Nieroda
Regional Computing
Center (RRZK)
University of Cologne,
Cologne, Germany
lnieroda@uni-koeln.de

Martin Peifer
Department of
Translational Genomics,
Center for Molecular
Medicine Cologne,
Medical Faculty
University of Cologne,
Cologne, Germany
mpeifer@uni-koeln.de

Viktor Achter
Regional Computing
Center (RRZK)
University of Cologne,
Cologne, Germany
vachter@uni-koeln.de

Janna Velder
Regional Computing
Center (RRZK)
University of Cologne,
Cologne, Germany
jvelder@uni-koeln.de

Ulrich Lang
Regional Computing
Center (RRZK)
University of Cologne,
Cologne, Germany
ulang@uni-koeln.de

ABSTRACT

In this paper we describe our design and experiences with the integration of iRODS with a partially automated pipeline, which was developed to optimize a cancer genomics workflow. In order to facilitate an efficient and secure data processing several challenges have to be overcome. The massive amount of data stemming from Next generation sequencing (NGS) requires a well structured data management system that allows basic storing and retrieval but also reviewing and sharing of both input as well as result data. It depends on the organization of the corresponding metadata whether it can be put to good use in subsequent analyses, thus we need to provide sufficient information for various levels – from a general overview to detailed comparisons of computation runs. The aspect of security and privacy is especially important when dealing with patient data, thus a fine grained approach to access authorization is an indispensable requirement. Finally, while High performance computing (HPC) systems provide the necessary compute power to perform high throughput data analysis, they also require careful handling to allow a robust workflow that can react accordingly to system or application errors. Our proposed solutions to the above mentioned aspects can be adapted to similar data processing workflows.

Keywords

Next generation sequencing, genome analysis, IRODS, workflow integration, High Performance Computing, Bioinformatics

INTRODUCTION

Next generation sequencing (NGS) is an increasingly cost efficient and reliable method to provide whole genomes or exomes (i.e., the protein coding part of the genome) in a relatively short time. Due to falling costs it became feasible to widen the scope of sequencing research and applications, for example from assembling a single human genome in the nineties, through analysis and comparison of thousands of genomes in the last decade up to the point where personalized medicine has been partially realized. The massive amounts of resulting data pose various challenges that need to be addressed in order to enable their exploration, analysis and effective dissemination. In particular,

iRODS UGM 2016 June 8-9, 2016, Chapel Hill, NC

genetic data runs through a lifecycle: the generated input is organized and stored depending on its type and origin, later on it is retrieved, processed and analyzed by high throughput machines in an HPC Cluster and finally, once the final results have been computed, they are made available for reviewing and further comparison. At each step the correct data needs to be identified, located, and processed in a secure way. Each time, a potential user access needs to be evaluated whether it is authorized and whether the proposed additions conform to the existing data schema or format. Those additions, e.g., quality control values or statistics regarding genome alignment are usually too extensive to be added to the file name and too valuable to be dumped into a text file without adequate means of search or aggregation. Traditional file systems quickly meet their limits both in terms of fine grained authorization as well as metadata, where content based information, e.g., about the project, sample ID, performed analysis and results are required. They operate with a simple username and -group principle and, if supported, with more advanced Access Control Lists (ACL) to manage access. The metadata is limited to system properties like ownership, size, date of creation or modification. It becomes apparent that a more sophisticated system is required.

As a computing center that has been driving NGS workflows for many years [1]-[2], we are constantly looking for solutions to meet such requirements and optimize these workflows to maximize output and quality. We have decided to use the comprehensive data management system IRODS, since it allows customized metadata attributes, fine grained protection rules as well as a query system to quickly organize and review the results. The integration in a HPC cluster provides the computing power necessary to perform high throughput data analysis framework. To facilitate a streamlined and robust workflow, we have automated the pipeline and introduced error handling procedures to react to both internal, content based as well as external, system based errors.

The described cancer genomics workflow or pipeline is an in-house development that has been successfully applied to a variety of large-scale genome sequencing projects [3, 4, 5, 6]. In the first step, the pipeline aligns raw whole genome or exome sequencing data to the reference genome using bwa-mem [7]. After alignment, the data is preprocessed to allow mutation detection. To this end, alignments are sorted, indexed, and potential PCR-duplications are masked. The difference to reference genome is determined in the next step together with quality control parameters of the sequencing run (e.g., mean coverage, insert size, etc.). Furthermore, genotype information as well as local read depths are extracted from the data. These derived data sets are the basis for mutation detection, where single nucleotides substitutions, small insertions and deletions, copy number changes, and genomic rearrangements (the latter only in case of whole genomes) are determined. Except for the alignment method, the entire set of computational methods are own developments that are uniformly implemented in C++ to allow an efficient processing of large genomic data sets.

Numerous NGS workflows have been adapted to HPC systems with various methods, e.g., HugeSeq [8] detects and annotates genetic variations by applying a MapReduce [9] approach, NGSANE [10] uses bash scripting with extensible logging and checkpointing measures, SIMPLEX [12] offers a fully functional VirtualBox Image to reduce installation issues. While they describe the analysis process in detail, few of them consider the requirements of data security and the necessary framework to make the results as well the corresponding metadata available for further dissemination. The WEP [11] pipeline for whole exome processing addresses the latter shortcoming by storing result metadata in a self developed MySQL database with a PHP-based web interface. What is missing is a comprehensive data management system that would encompass the employed input data, the results and the metadata within a secure and reliable framework. Even though the metadata delivers necessary information, the underlying files should also be stored in a controlled environment, so that they can be both retrieved at a moments notice.

There are organizations that have employed iRODS for their NGS workflows, namely the Wellcome Trust Sanger Institute [13], Broad Institute, Genome Center at Washington University, Bayer HealthCare, University of Uppsala and others, but to our knowledge only the Sanger Institute has published their experiences in a peer-reviewed journal so far. However, they restrict the use of iRODS to store and manage alignment files (in BAM format) only.

In contrast, by merging our workflow with iRODS, we could not only store and annotate the input data with relevant information but also parse the results and make them available for queries through self defined metadata within a single system.

IRODS

We have based our data management system implementation on iRODS due to several reasons. The most prominent feature is the possibility to add customized metadata to all stored files, thus allowing researchers to track and manage their original input and results. This includes descriptions of sample origins, various types of performed tests and analysis runs as well as their actual result values that can be subsequently searched via SQL like queries. It also provides a rule-engine that enables the execution of predefined actions in regular intervals, triggered by certain events or manual control. Any data operations can thus be augmented with matching actions, e.g., leaving an audit trail, submitting HPC jobs once certain data files arrive or sending a message after particular events. IRODS uses a virtual file hierarchy that can be adapted to various organizational structures and is independent from the actual physical storage. The access control can be easily fine tuned from encompassing groups down to single users, giving a flexibility similar to POSIX ACLs, regardless whether the underlying file system supports it.

IMPLEMENTATION

The CHEOPS Cluster is an HPC Cluster that provides processing power of 100 Teraflop/s Peak (85,9 Teraflop/s Linpack) for scientists within the german state of North Rhine-Westfalia. It is comprised of 841 nodes using Nehalem, Nehalem EX and Westmere Intel architectures. The primary interconnect is realized with Infiniband QDR hardware, ensuring a low latency network with 40Gb/s bandwidth. A secondary network dedicated to managerial and NFS functionality is built with 10 Gigabit Ethernet (10GbE). The storage system relies on a GridScaler for archival purposes and Lustre for fast, temporary computations.

We have installed the IRODS ICAT server version 4.1.8 together with a MySQL database on a virtual machine (VM) which is connected to CHEOPS and a few research institutes, among them CECAD, via 10 Gigabit Ethernet. The actual bandwidth may be lower, due to shared connections with other VMs hosted by the ESX-server.

Security concerns

Since genome sequencing provides the complete genomic fingerprint of the patient, high security standards are an indispensable requirement while handling of such data. To protect data transmissions en route as well as to restrict access to predefined hosts, all IRODS iCommand clients have been set up to communicate through host-certificate based SSL encryption. When we tested version 4.1.6, the kerberos plugin did not seem to be compatible with the rest of the distribution and we have thus resorted to PAM based authentication. The tests have not been repeated for 4.1.8 and remain an option for future work. The IRODS Server has a main vault directory, which holds the data archive and is owned exclusively by the `irods` user. It is physically located on the GridScaler Storage System and made available to iRODS via an NFS mount, which can also be accessed only by the `irods` user. Only the VM provides iRODS resources, all other machines have to use iCommand Clients or APIs to down-, upload and query the data. With such a setup any data located in the vault is shielded from all CHEOPS users through locally managed file ownership and permissions but the security within the vault depends entirely on the strength and infallibility of iRODS authorization mechanisms. A plausible alternative was to use iRODS' registering feature, which can register objects in the database based on their physical storage path and allows to create respective metadata associations. While such an approach would remove the reliance on iRODS' internal security, it would also present several potential problems. It relays file management entirely to the underlying file system, which has to support sufficiently fine grained access control. The registering operation can be made either by the administrator or by the file owner which requires loosening of iRODS permission checking rules as well as NFS export restrictions, which leads to further concerns. The direct access plugin raises similar concerns, as it requires super user access. Furthermore, by ceding any control over the files, iRODS information is no longer reliable and can point to non existent or, even worse, to altered or replaced files with different contents. Instead, we have decided to let iRODS maintain its data within the vault but restrict it to a dedicated server with stringent access policies.

Data consistency

To ensure metadata consistency both content- as well as format-wise, we have defined a simple schema for the data import and run execution steps. The data provider preparing raw input data or, respectively, the scientist preparing an analysis run fill out a metadata sheet with predefined attributes that must adhere to corresponding value domains and check procedures. Those are parsed by perl scripts, which validate them and perform additional tests, if required by the schema. See an excerpt in Table 1. The attributes depend on the step type – the import sheets are designed to describe the data origin as thoroughly as possible while the run sheets focus on their application, e.g., the reference to align the data against. Some of the attributes are generated by the scripts, depending on encountered file format, creation date and other factors.

Attribute	Example value	Value domain	Further tests
LocalPath	/projects/username/sample	[a-zA-Z0-9_-+/.]	Path readability
Filename	testfile_T.bam	[a-zA-Z0-9_-+/.].{bam|fastq}	File readability
Sample_ID	P1234-PB03	[a-zA-Z0-9_-+.]	None
Data Provider	Max Mustermann	[a-zA-Z-]	None

Table 1. Excerpt from import schema

The perl scripts create Virtual Paths in iRODS according to certain attribute values, like the Projectname or Sample_ID, to place and subsequently find the data in unique, predefined locations which have been cleared for the submitting user or his group, e.g.

```
/<Zone>/archive/<Projectname>/<Sample_Type>/<Sample_ID>/input/
/<Zone>/archive/<Projectname>/<Sample_Type>/<Sample_ID>/run_1/
/<Zone>/archive/<Projectname>/<Sample_Type>/<Sample_ID>/run_2/
...
```

The upload can only succeed if the user has sufficient rights for the Zone, Project and Sample_Type and thus the generated Virtual Path. Once the metadata has been validated, tested and extended with dynamically generated content, it is packed into Attribute-Value-Unit (AVU) triplets and sent together with the input files via `iput` command.

Use Case Scenario

The pipeline for sequencing and analysing cancer exomes and genomes is comprised of several steps. See Figure 1 for an overview of the relevant data flow. The next-generation sequencer generates short snippets of genetic sequences or "reads" in FastQ format, which are subsequently converted into the binary BAM format and stored in the institute's local storage. An employee prepares a corresponding metadata input sheet and launches the "import" script which parses, validates, tests and extends it with dynamically generated content. It organizes the metadata into Attribute-Value-Unit (AVU) triplets and uploads it together with the BAM files to the iRODS server via `iput` command. The server tags the files, changes their ownership to the `irods` user and stores them in its vault directory on the GridScaler, which it can access through an NFS mount. At the end of this step the input files are archived on the CHEOPS cluster and associated with their respective descriptions.

For the analysis process, the metadata run sheet is prepared and launched with a "run" script, which validates it and executes the remainder of the pipeline. First, it uses the parsed attributes, like Sample_ID, Projectname and Reference to locate and download the appropriate files from predefined paths. This is implemented with simple iRODS rules, which are executed with the `irule` command. Below is a short example for retrieving samples files that match given criteria.

```
getSampleFiles {
```

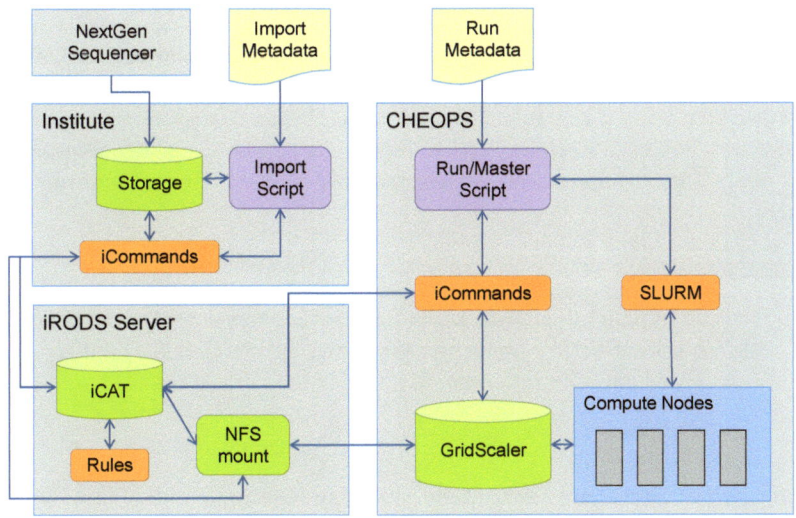

Figure 1. Data flow diagram

```
#Input parameters:
#   Sample_ID, Sample_Type, Project, Zone
#Output:
#   List of Matching input files
#Example launch:
# irule -F getSampleFilesExt.r   "*Sid='testid'" "*Project='testproject'"
 *Coll="/*Zone/archiv/*Project/*Sid/*Stype/input"
 msiExecStrCondQuery("SELECT DATA_NAME WHERE COLL_NAME = '*Coll' and \
     META_DATA_ATTR_NAME = 'Sample_ID' and META_DATA_ATTR_VALUE = '*Sid'",*QOut);
 foreach(*QOut) {
   msiGetValByKey(*QOut,"DATA_NAME", *File);
   writeLine("stdout","*Coll/*File");
 }
}
INPUT *Sid="test", *Stype="exome", *Project="test", *Zone="SMOOSEzone"
OUTPUT ruleExecOut
```

With their locations known, they are downloaded with the `iget` command and the actual data analysis is started. This process is divided into four distinct phases and during each one the script generates jobscripts for the SLURM scheduling system, submits the jobs to the cluster, waits for their execution and depending on the outcome proceeds to the next phase. Should an error be encountered, whether it is a failed job or inconsistent results, e.g., non matching read numbers from the original and annotated BAM files, the pipeline is aborted with an appropriate error message. Once the analysis is finished, the results including statistics, quality control and relevant logs are parsed and packed into AVU triplets. As a final step, the script uploads the data and corresponding metadata to the iRODS server, which archives it and makes it available for further analysis and dissemination.

CONCLUSION

The described pipeline automation and integration with iRODS empowers scientists to keep track of their data in an efficient and secure manner. By employing verifiable data schemas, we could enforce metadata consistency and build a hierarchical structure within iRODS' virtual file space that placed files in predefined locations. While it provided a straightforward means to narrow data searches down, it also made mapping of user permissions easier to manage. The possibility to restrict access to certain projects or file groups is especially relevant in the clinical context, where patient data is involved. We have decided to rely on IRODS' authentication in order to let it manage contents in their entirety, rather than using it as a sole metadata provider. For this means we have also tightened security and restricted its services to a virtual machine. The inclusion of both the input as well as output data with matching descriptions has resulted in a comprehensive system that allows to retrieve and compare analysis results with their underlying sources.

ACKNOWLEDGEMENTS

We would like to thank Mr. Carsten Jahn from Bayer Business Services GmbH, HealthCare Research who has shared his experiences with using iRODS. This work was supported by the German Ministry of Science and Education (BMBF) as part of the e:Med initiative (grant no. 01ZX1303A).

REFERENCES

[1] Achter Viktor, Seifert Marc, Lang Ulrich, Götze Joachim, Reuther Bernd, Müller:Nachhaltigkeitsstrategien bei der Entwicklung eines Lernportals im D-Grid, Lecture Notes in Informatics (LNI). Proceedings Vol P-149: 43–54, 2009

[2] Kawalia Amit, Motameny Susanne, Wonczak Stephan, Thiele Holger, Nieroda Lech, Jabbari KAmel, Borowski Stefan, Sinha Vishal, Gunia Wilfried, Lang Ulrich, Achter Viktor, Nürnberg Peter: Leveraging the Power of High Performance Computing for Next Generation Sequencing Data Analysis: Tricks and Twists from a High Throughput Exome Workflow. PLoS One, 2015

[3] Peifer M., Hertwig F., Roels F., Dreidax D., Gartlgruber M., Menon R., Krämer A., et al.: Telomerase activation by genomic rearrangements in high-risk neuroblastoma. Nature 526:700-704, 2015

[4] George J., Lim J.S., Jang S.J., Cun Y., Ozretic L., Kong G., Leenders F., et al.: Comprehensive genomic profiles of small cell lung cancer. Nature 524:47-53, 2015

[5] Fernández-Cuesta L., Peifer M., Lu X., Sun R., Seidel D., Zander T., Leenders F., et al.: Frequent mutations affecting chromatin remodeling genes in pulmonary carcinoids. Nature Communications 5:3518, 2014

[6] Peifer M., Fernández-Cuesta L., Sos M.L., George J., Seidel D., Kasper L.H., Plenker D., et al.: Integrative genome analyses identify key somatic driver mutations of small-cell lung cancer. Nature Genetics 44:1104–1110, 2012

[7] Li H.: Toward better understanding of artifacts in variant calling from high-coverage samples. Bioinformatics 30:2843-2851, 2014

[8] Lam Hugo Y. K., Pan Cuiping, Clark Michael J., Lacroute Phil, Chen Rui, Haraksingh Rajini, O'Huallachain Maeve, et al.: Detecting and annotating genetic variations using the HugeSeq pipeline. Nature Biotechnology 30(3): 226–229, 2012

[9] Dean J., Ghemawat S.: MapReduce: simplified data processing on large clusters. in OSDI'04 Proceedings of the 6th Symposium on Operating Systems Design and Implementation, San Francisco, 2004.

[10] Buske Fabian A., French Hugh J., Smith Martin A., Clark Susan J., Bauer Denis C.: NGSANE: A lightweight production informatics framework for high-throughput data analysis. Bioinformatics. 30(10): 1471–1472, 2014

[11] D'Antonio Mattia, D'Onorio De Meo Paolo, Paoletti Daniele, Elmi Beradino, Pallocca Matteo, Sanna Nico, Picardi Ernesto, Pesole Graziano, Castrignano Tiziana: WEP: a high-performance analysis pipeline for whole-exome data. BMC Bioinformatics, 14(Suppl 7): S11, 2013

[12] Fischer Maria, Snaider Rene, Pabinger Stephan, Dander Andreas, Schossig Anna, Zschocke Johannes, Trajanowski Zlatko, Stocker Gernot: SIMPLEX: Cloud-Enabled Pipeline for the Comprehensive Analysis of Exome Sequencing Data. PLoS One, 2012

[13] Chiang Gen-Tao, Clapham Peter, Qi Guoying, Sale Kevin, Coates Guy: Implementing a genomic data management system using iRODS in the Wellcome Trust Sanger Institute. BMC Bioinformatics, 12:361, 2011

Status and Prospects of Kanki: An Open Source Cross-Platform Native iRODS Client Application

Ilari Korhonen,* Miika Nurminen
IT Services, University of Jyväskylä
PO Box 35, 40014 University of Jyväskylä, Finland
ilari.korhonen@icloud.com, miika.nurminen@jyu.fi

ABSTRACT

The current state of development of project Kanki is discussed and some prospects for future development are laid out with reflection on the results of the research IT infrastructure project at the University of Jyväskylä. Kanki is a cross-platform native iRODS client application which was introduced to the iRODS community at the iRODS Users Group Meeting in 2015, and later released as open source. A total of 9 releases have been made, from which the latest 6 have been available in addition to the source code as pre-built binary packages for x86-64 CentOS Linux 6/7 and OS X 10.10+. The Kanki build environment at the University of Jyväskylä is running out of Jenkins continuous integration for both previously mentioned platforms utilizing disposable containers instantiated from pre-built Docker images for Linux builds. This provides an excellent framework for (regression) testing of the client suite. The immediate goals of development include: stability, testing, ease of install and use, and a complete iRODS basic feature set for graphical icommands alternatives. The prospects for more advanced future development include: a fully extensible modular metadata editor with pluggable attribute editor widgets, a fully extensible modular search user interface with pluggable condition widgets, data grid analytics, and visualization with VTK integration.

Keywords

Research data, iRODS, client software, graphical user interface, continuous integration, research support services.

INTRODUCTION

About a year ago in 2015, the cross-platform native iRODS client application Kanki was introduced to the iRODS community at the 7th Annual iRODS Users Group Meeting [1]. The Kanki iRODS client features a responsive UI with native look & feel to the desktop enabled by the Qt^1 framework, integration to Kerberos authentication with the option to use iRODS 4.x SSL secured connections. Metadata management features include a schema definition language, field validators, and type-specific display views and filtering. The client application is targeted towards researchers of various disciplines as well as other interest groups utilizing or curating research data (e.g. librarians). Users can utilize the full power of an iRODS data grid complete with powerful data management functions via its intuitive user interface. Kanki was released as open source with a 3-clause BSD license in GitHub[2] in September 2015. Kanki has been used in pilot projects at the University of Jyväskylä and has attracted interest from other research institutions as well.

In this paper, we focus on the recent development efforts of the Kanki iRODS client, as well as reflection on the research IT infrastructure development project that enabled the development. The paper is concluded with development ideas and a possible sketch on the role of Kanki and iRODS as part of the research support services at the JYU.

---

*Present address: PDC Center for High Performance Computing, KTH Royal Institute of Technology, SE-100 44, Stockholm, Sweden

[1] http://www.qt.io/

[2] https://github.com/jyukopla/kanki-irodsclient

iRODS UGM 2016 June 8-9, 2016, Chapel Hill, NC

KANKI DEVELOPMENT

Kanki is being developed as a cross-platform project with a single C++11 code base. Currently Kanki builds successfully on Linux (Red Hat and Ubuntu tested and documented) and Mac OS X 10.10/10.11. The multi-platform portability of the source is enabled by the portability of the Qt framework and the C++11 standard library.

Build Process

Currently the multi-platform build is being done via the Qt **qmake** utility, which generates a build environment to be executed with GNU Make. There is a simple build script in the GitHub repository called **build.sh** which builds the source package on Linux and OS X. The build script can be provided the location of the Qt framework as well as some other arguments. Help can be printed out with the **-h** switch.

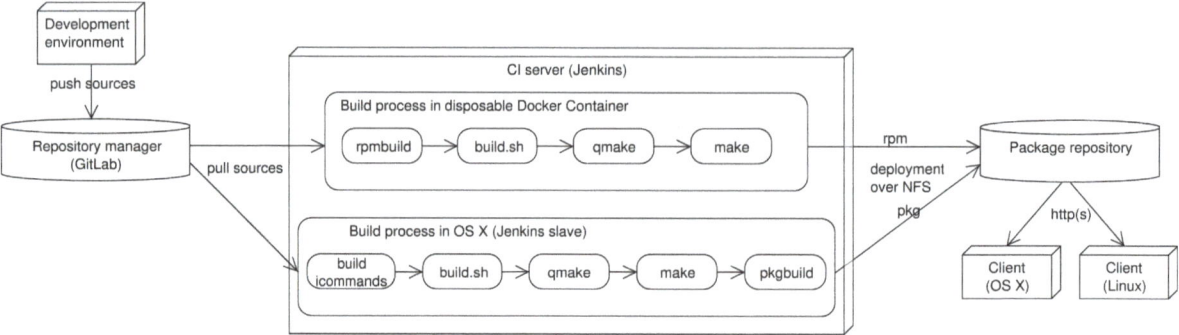

Figure 1. Illustration of the kanki-irodsclient build process.

Since the iRODS project is migrating its build environment into CMake (which also enables very convenient VTK/Qt linkage), work is currently in progress to switch the build environment to CMake. There is an experimental build environment which builds via CMake in the **develop** branch of the GitHub repository. A shadow build via CMake can be done as follows:

```
$ git clone https://github.com/ilarik/kanki-irodsclient.git -b develop
$ mkdir build_kanki-irodsclient; cd build_kanki-irodsclient
$ cmake -DCMAKE_PREFIX_PATH:PATH=/Users/tiilkorh/Qt/5.5/clang_64/lib/cmake ../kanki-irodsclient/src
-- The C compiler identification is AppleClang 7.3.0.7030031
-- The CXX compiler identification is AppleClang 7.3.0.7030031
-- Check for working C compiler: /Applications/Xcode.app/Contents/Developer/Toolchains/XcodeDefault.xctoolchain/usr/bin/cc
-- Check for working C compiler: /Applications/Xcode.app/Contents/Developer/Toolchains/XcodeDefault.xctoolchain/usr/bin/cc -- works
-- Detecting C compiler ABI info
-- Detecting C compiler ABI info - done
-- Detecting C compile features
-- Detecting C compile features - done
-- Check for working CXX compiler: /Applications/Xcode.app/Contents/Developer/Toolchains/XcodeDefault.xctoolchain/usr/bin/c++
-- Check for working CXX compiler: /Applications/Xcode.app/Contents/Developer/Toolchains/XcodeDefault.xctoolchain/usr/bin/c++ -- works
-- Detecting CXX compiler ABI info
-- Detecting CXX compiler ABI info - done
-- Detecting CXX compile features
-- Detecting CXX compile features - done
-- Configuring done
-- Generating done
-- Build files have been written to: /Users/tiilkorh/tmp2/build_kanki-irodsclient
$ make -j 8
[ ... lots of nicely formatted CMake build output ... ]
[100%] Linking CXX executable irodsclient.app/Contents/MacOS/irodsclient
[100%] Built target irodsclient
```

The build environment for the Kanki iRODS client at JYU is running out of Jenkins CI on two slaves, one with Docker capabilities and another slave running OS X. Linux builds are currently being executed in disposable containers instantiated from pre-built Docker images in a Jenkins slave with Docker. The OS X builds are being done against a prebuilt (another Jenkins job) irods-icommands distribution for OS X from the **4-1-stable** branch of the iRODS GitHub repository. Instructions for the OS X build have been published in the iRODS blog[3].

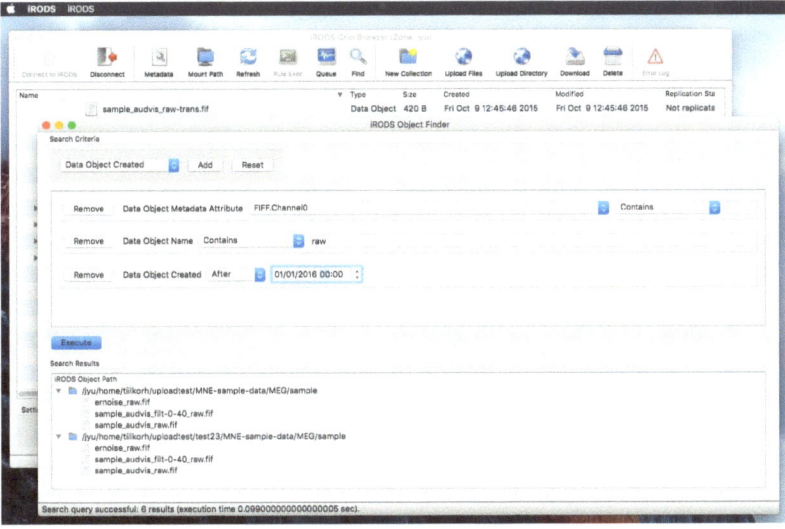

Figure 2. Object Finder component.

New and Prospective Features

Specific needs not properly accommodated by other existing freely available solutions included the graphical iRODS search tool with arbitrary search criteria formation for data discovery, metadata schema management with visual namespace and attribute views, and readiness for metadata schema validation for data quality assurance.

As the client is intended to eventually serve as a bona fide alternative user interface to iRODS icommands – the reference user interface for iRODS. Implementing all of this functionality in a native "desktop" application will enable the users to harness the full power of iRODS with native application performance and easy-to-learn graphical user interface. The following features are considered to be developed further:

- drag & drop inside iRODS and between desktop and iRODS

- synchronization of iRODS collections with local paths

- editing of access control lists and groups for groupAdmin role users

- metadata editor schema management with validation

- rule engine queue management and rule exec interface

- data discovery i.e "Find" UI for arbitrary metadata search criteria execution (GenQuery) (see Figure 2)

- VTK[4]-based visualization tools for data grid analytics (e.g. object relations, see Figure 3)

---

[3] http://irods.org/2015/10/native-gui-access-to-irods-on-a-mac-or-linux-desktop/
[4] http://www.vtk.org/

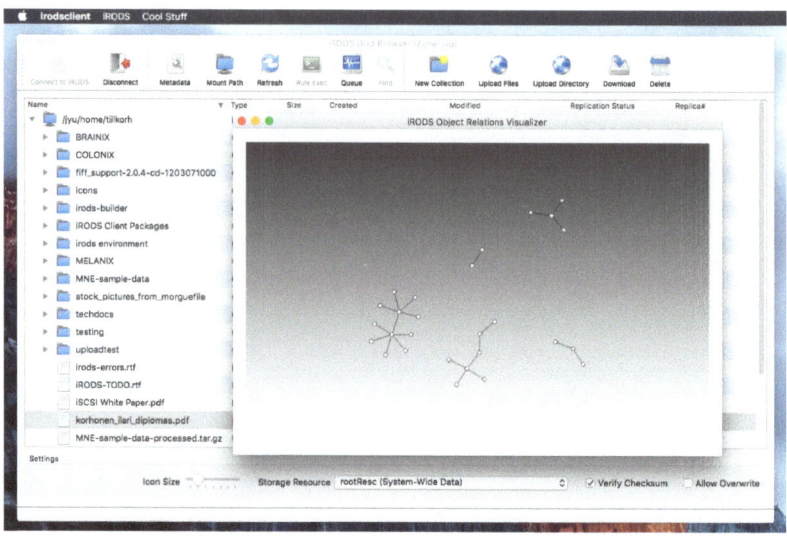

Figure 3. Experimental iRODS Object Relations Visualizer.

Windows still remains as an unsupported platform since iRODS 4.0 isn't Windows compatible at the time of writing this paper. With Windows support added to the iRODS codebase our client can be built on Windows as well.

REFLECTION ON THE RESEARCH IT INFRASTRUCTURE DEVELOPMENT PROJECT

Initial implementation of Kanki client has been done under the development project for research IT infrastructure and research data management, active in 2013-2015 at the University of Jyväskylä. One of the project goals was putting the university-level principles for research data management [2] (e.g. processes for handling research data, secure infrastructure for data storage and access, standard metadata descriptions, advancing open science) into practice. Dataverse[5] was first adopted, making it possible to publish datasets in citable format with on-line analysis capability [3]. iRODS data grid has been adopted with additional development of server-side iRODS modules for e.g. automatic metadata extraction and finally, the development of Kanki client. A recommended minimum metadata model has been developed with JYU Library, providing also training for research data management practices.

iRODS and Kanki client have been utilized in several pilot projects, including the Jyväskylä Centre for Interdisciplinary Brain Research[6], JYFL Accelerator laboratory[7], and datasets from the Department of Music. Various datasets – including video, audio, motion detection data, EEG measurements, and transcribed interviews – have been imported to the system. Metadata extraction functionality has been implemented for DICOM[8], EXIF[9], IPTC[10], and FIFF[11] formats. Faculty of Information Technology is evaluating the suitability of iRODS as the storage backend for Faculty's local computing cluster.

There have been multiple, partly concurrent efforts related to research data management and, more generally, research support services – most of these overseen at the JYU by Steering group for Research information systems[12], Steering

---

[5] https://dvn.jyu.fi/

[6] http://cibr.jyu.fi/

[7] https://www.jyu.fi/fysiikka/en/research/accelerator/

[8] http://dicom.nema.org/standard.html

[9] http://www.jeita.or.jp/cgi-bin/standard_e/list.cgi?cateid=1&subcateid=4

[10] https://iptc.org/standards/photo-metadata/iptc-standard/

[11] http://www.aston.ac.uk/lhs/research/centres-facilities/brain-centre/facilities-clinical-services/meg-studies/downloads/

[12] https://www.jyu.fi/hallinto/tyoryhmat/tutkimuksen_tietojarjestelmat_or/en

group for the Digitalisation project[13], and the Science council[14]. Nationally, Open Science and Research (ATT[15]) project has produced services and recommendations related to e.g. open research data. All departments of the University have provided documentation of their research infrastructures[16]. Procedures for digitising analogue content have been implemented by the University Printing Services. JYU is currently replacing the legacy research information system[17] with a new CRIS[18] (Current Research Information System). Jyväskylä University Library is planning a model for centralized open science services and promoting parallel publishing in a joint development project with University of Eastern Finland Library [4], expanding on highly successful centralized publication recording process [5] in JYU, realized by close collaboration between the Library, Research management, and IT Services. National Data Management Plan tool Tuuli[19] is being tested by various Finnish universities. After the adoption of the new CRIS, data management plans will be systematically collected as part of research project -specific records.

Despite the various efforts and activities in the IT Research infrastructure project, it has to be concluded that even though the implementation of Kanki client and the technical infrastructure can be considered successful, the wider goal of executing the principles of data management has turned out to be challenging. It has proved to be of considerable difficulty to advance standard data management practices when the research itself is done more or less independently of administrative processes, often using specialized tools and software for e.g. analysis on datasets. The majority of JYU researchers have not yet used iRODS or Dataverse. IT Services unit has sometimes been perceived as not sufficiently accommodating the needs of various research groups that may have differing requirements in terms of hardware, software, scalability, or access rights. The services have been improved during the project, but yet more steering may be needed related to job function scoping and e.g. pricing models – the cost of centralized storage has not been competitive compared to ad hoc solutions such as portable drives. Another identified challenge has been the lack of resources related to coordinating the activities of different actors related to research support services.

CONCLUSION

The development of Kanki client is still at relatively early stages. The solution has provoked interest from multiple research institutions and developer community seems to be building up in GitHub. The build process utilizing Jenkins and Docker seems promising considering regression testing and streamlined deployment, possibly utilizing Ansible in the server-side in the future. iRODS maintenance at JYU is being transferred from IT development services to storage services, thus making the iRODS environment as part of the standard service porfolio. Development of Kanki client continues as open source project. The software can be considered adequate in production settings, but is by no means "ready". Immediate goals of development include stability, testing, ease of install and use, and a feature set for graphical icommands alternatives. Concerning the usage related to different stages of research data lifecycle, longer-term development prospects include: a fully extensible modular metadata editor with pluggable attribute editor widgets, a fully extensible modular search user interface with pluggable condition widgets, data grid analytics, and additional VTK-enabled visualizations.

Even though iRODS or Dataverse have not yet been widely adopted by JYU researchers, the situation might change in the near future. It is likely that as funders like Academy of Finland or EU (Horizon 2020) start to demand open research data, repository services – be it university-specific, international (e.g. Zenodo, FigShare), or subject-based (e.g. Finnish Social Science Data Archive[20] – shall become more attractive. The general data protection regulation[21] in EU will considerably raise the requirements for all systems containing personal data – this will likely generate new use cases for iRODS.

---

[13]https://www.jyu.fi/hallinto/tyoryhmat/digitointityoryhma

[14]https://www.jyu.fi/hallinto/neuvostot/tiedeneuvosto/en/sciencecouncil

[15]http://openscience.fi/

[16]https://www.jyu.fi/palvelut/wolmar/palvelut/infrat

[17]http://tutka.jyu.fi/

[18]https://www.jyu.fi/yliopistopalvelut/str/erityistoiminnot/tietohallinto/cris-kayttoonottoprojekti

[19]https://www.dmptuuli.fi/

[20]http://www.fsd.uta.fi/en/

[21]http://www.consilium.europa.eu/en/policies/data-protection-reform/data-protection-regulation/

JYU Library and University Museum will form a new unit named *Open Science Centre* (OSC) as part of a more extensive organizational transformation – the University's structural development[22] – effective in 2017. The University is also mapping the present state of research support services and identifying potential development needs and gaps wrt. National Open Science and Research Reference Architecture[23]. The new organization might provide opportunities to further clarify service models and distribution of responsibilities – perhaps even establishment of new centralized functions. For example, there is an emerging need for new services to support open research *methods* (e.g. source code and notebooks [6]), and to perceive an expanding variety of tools related to different kinds or research workflows [7]. The long term goal for research IT infrastructure (at least for data-intensive disciplines) could be supporting fully reproducible research [8]: making the code and data available in a platform such that the data can be analyzed in a similar manner as in the original publication.

In the new organization, OSC is to take more comprehensive responsibility on coordinating research data management and digitisation activities. However, execution of the principles requires still substantial technical development, as well as architectural steering. CRIS has potential to be used as a (meta)data hub combining information about research infrastructures, projects, and outputs (e.g. publications and datasets), provided that sufficient resources are reserved for data integration. There is also a partial overlap between functionalities of the Institutional repository[24] and Dataverse. Further opportunities for integrating data or even establishing shared systems (cf. Tuuli DMP, National ORCID consortium[25]) by multiple universities should be explored to minimize duplicated work [9]. An important issue for the future is the acceptance of iRODS by the researchers. There is a growing need to support data management during the research life cycle as a whole [10] – we believe that iRODS is the principal enabling service to accomplish this, provided that it remains supported with sufficient advocacy and training.

REFERENCES

[1] I. Korhonen and M. Nurminen, "Development of a native cross-platform iRODS GUI client," in *Proceedings of iRODS User Group Meeting 2015*. The iRODS Consortium, 2015.

[2] A. Auer and S.-L. Korppi-Tommola, "Principles for research data management at the University of Jyväskylä," University of Jyväskylä, Tech. Rep., 2014. [Online]. Available: https://www.jyu.fi/tutkimus/tutkimusaineistot/rdmenpdf

[3] M. Crosas, "The Dataverse network®: an open-source application for sharing, discovering and preserving data," *D-Lib Magazine*, vol. 17, no. 1, 2011.

[4] P. Olsbo, A. Muhonen, J. Kananen, and J. Saarti, "Suomi rinnakkaistallentamisen mallimaaksi [Finland to become a model country for parallel publishing]," 2015, Portti. [Online]. Available: http://portti.avointiede.fi/tutkimusjulkaisut/suomi-rinnakkaistallentamisen-mallimaaksi

[5] M.-L. Harjuniemi, "Kirjasto yliopiston julkaisurekisterin ylläpitäjänä: kokemuksia keskitetystä kirjaamisesta [Library as maintainer of university's publication registry: experiences of centralized recording]," *Signum*, no. 2, 2015.

[6] A. Perrier, "Jupyter, Zeppelin, Beaker: The rise of the notebooks," 2015, ODSC blog. [Online]. Available: https://www.opendatascience.com/blog/jupyter-zeppelin-beaker-the-rise-of-the-notebooks/

[7] B. Kramer and J. Bosman, "101 innovations in scholarly communication - the changing research workflow," 2015, poster at FORCE2015 Conference.

[8] V. Stodden, F. Leisch, and R. D. Peng, Eds., *Implementing Reproducible Research*. Chapman&Hall/CRC,2014.

[9] M. Nurminen, "Preparing for CRIS: Challenges and opportunities for systems integration at Finnish universities," 2014, poster at Open Repositories 2014.

[10] T. Walters, "Assimilating digital repositories into the active research process," in *Research Data Management – Practical Strategies for Information Professionals*, J. M. Ray, Ed. Purdue University Press, 2014.

---

[22] https://www.jyu.fi/hallinto/strategia/strategiat/strategian_toimenpideohjelma_en
[23] https://avointiede.fi/viitearkkitehtuuri
[24] https://jyx.jyu.fi/
[25] https://tutkijatunniste.fi/